Computational Molecular Science and Engineering Forum

Presentations at the 2009 AIChE Annual Meeting

Nashville, Tennessee, USA
8-13 November 2009

ISBN: 978-1-61567-942-3

Printed from e-media with permission by:

Curran Associates, Inc.
57 Morehouse Lane
Red Hook, NY 12571

Some format issues inherent in the e-media version may also appear in this print version.

Printed by Curran Associates, Inc. (2010)

For permission requests, please contact AIChE
at the address below.

AIChE
3 Park Avenue
New York, NY 10016-5991

Phone: (203) 702-7660
Fax: (203) 775-5177

www.aiche.org

Additional copies of this publication are available from:

Curran Associates, Inc.
57 Morehouse Lane
Red Hook, NY 12571 USA
Phone: 845-758-0400
Fax: 845-758-2634
Email: curran@proceedings.com
Web: www.proceedings.com

TABLE OF CONTENTS

Coarse-Grained Lattice Monte Carlo Simulations of Continuous Systems

Xiao Liu, Department of Chemical and Biomolecular Engineering, University of Pennsylvania, Philadelphia, PA

Warren D. Seider, Chemical and Biomolecular Engineering, University of Pennsylvania, Philadelphia, PA

Talid Sinno, Chemical and Biomolecular Engineering, University of Pennsylvania, Philadelphia, PA

Monte Carlo and its dynamic variant, kinetic Monte Carlo, are used extensively to simulate an enormous range of material properties. In both cases, restricting particle positions to fixed lattice sites can substantially increase the computational efficiency of a simulation, and this benefit increases as the lattice becomes coarser. However, the confinement of particle positions to a rigid lattice necessarily reduces the available configurational degrees of freedom in a system and this constraint can become very important at elevated temperatures [1,2].

In this presentation, we discuss a new technique for performing Metropolis Monte Carlo (MMC) simulations of continuous systems on coarse rigid lattices, while preserving the phase-space contributions of the missing degrees-of-freedom. Previous Metropolis Monte Carlo modeling studies have shown that lattice representations are able to correctly capture the phase behavior of continuous fluids if the discretization is sufficiently fine [3], but the potential computational gains are limited by the required lattice resolution. The present approach relies on the pre-computation of coarse-grained potentials from equilibrium sampling of small systems. The potentials are generated in such a way so as to be scalable to different temperatures without the need for a new set of calculations. It is shown that the approach is able to capture both equilibrium (such as phase diagram features) and non-equilibrium (transport) features of the continuous system, even when the rigid lattice becomes highly coarse.

The approach then is extended to lattice kinetic Monte Carlo (LKMC) simulations by developing a method to map particle motions in a continuous system onto a finite set of discrete rate expressions. Both accuracy and performance are discussed with examples based on the Lennard-Jones potential.

[1] S. S. Kapur, M. Prasad, J. C. Crocker, T. Sinno, Role of configurational entropy in the thermo-dynamics of clusters of point defects in crystalline solids, Phys. Rev. B 72, 014119 (2005).

[2] J. Dai, W. D. Seider and T. Sinno, Lattice kinetic Monte Carlo simulations of defect evolution in crystals at elevated temperature. Molecular Simulation, 32, 305 (2006).

[3] A. Z. Panagiotopoulos, On the equivalence of continuum and lattice models for fluids. J. Chem. Phys., 112, 7132 (2000).

See more of this Session: Recent Advances in Molecular Simulation Methods I
See more of this Group/Topical: Computational Molecular Science and Engineering Forum

Fast Lattice Monte Carlo Simulations of Soft Materials

Qiang Wang, Chemical and Biological Engineering, Colorado State University, Fort Collins, CO

We introduce fast lattice Monte Carlo (FLMC) simulations, where multiple occupancy of lattice sites is allowed with a proper Boltzmann weight and thus the evaluation of nearest-neighbor interactions can also be avoided. FLMC simulations (with multiple occupancy of lattice sites and Kronecker δ-function interactions) are much more efficient, in the study of equilibrium properties of soft materials such as polymers, than both conventional lattice MC simulations (with self- and mutual-avoiding walks and nearest-neighbor interactions) and fast off-lattice MC simulations (with pair-potential calculations). When compared with the corresponding lattice field theories based on the *same* Hamiltonian, FLMC simulations further provide a powerful means for unambiguously and quantitatively revealing the effects of long-wavelength correlations/fluctuations. Using confined homopolymers as an example, we report for the first time FLMC simulation data ranging from the single-chain case all the way to the lattice self-consistent field limit.

See more of this Session: Recent Advances in Molecular Simulation Methods I
See more of this Group/Topical: Computational Molecular Science and Engineering Forum

Direct Simulation Approach for Computing the Interfacial Free Energy for Nucleation From Solution

Nathan Duff, Chemical Engineering, University of California, Santa Barbara, Santa Barbara, CA

Baron Peters, Chemical Engineering, University of California, Santa Barbara, Santa Barbara, CA

Nucleation from solution is important in many pharmaceutical crystallization processes, biomineralization, materials synthesis, and self-assembly[1]. Simulation methodology has progressed rapidly for nucleation in pure-component and implicit solvent systems; however there has been relatively little progress for simulating nucleation from a solution with explicit solvent molecules. The impasse stems from problems in combining the specialized methods for simulating rare barrier-crossing events, with specialized particle insertion algorithms for maintaining a constant chemical potential driving force (supersaturation) during nucleation[2]. At present there are no algorithms that simultaneously address these two central aspects of nucleation from solution. We present a lattice gas - Potts model (LGP)[3] to aid in the development of new simulation strategies for nucleation from solution. To our knowledge the LGP is the first general lattice model to capture common crystallization phase diagram features like a eutectic point and solute/solvent melting points. Simulations of nucleation in the LGP reveal a competition between nucleation of amorphous and crystalline particles. A simple explanation for this commonly observed phenomenon is developed from classical nucleation theory. Finally, we demonstrate a new "mitosis" simulation method for directly computing the interfacial free energy of a nucleus with the solution. The method entirely avoids the need for particle insertion or particle identity changes, so it should be applicable for atomistic simulations of real crystal nucleation processes in solution. The new approach should provide a path forward for using simulations to help engineer solvent additives to direct polymorph selection and self-assembly.

1. Lovette MA, Browning AR, Griffin DW, Sizemore JP, Snyder RC, Doherty MF. Crystal Shape Engineering. *Industrial & Engineering Chemistry Research.* 2008;47(24):9812-9833.

2. Frenkel D, Smit B. *Understanding Molecular Simulation*. 2nd ed. London: Academic Press, Inc.; 2002.

3. Chaikin PM, Lubensky TC. *Principles of condensed matter physics*. New York: Cambridge University Press; 2000.

See more of this Session: Recent Advances in Molecular Simulation Methods I
See more of this Group/Topical: Computational Molecular Science and Engineering Forum

Simulations of Phase Transitions, Metastability, and Nucleation Via Spatial Updating and Tempering Techniques

G. Orkoulas, Chemical and Biomolecular Engineering, University of California at Los Angeles, Los Angeles, CA

Spatial updating grand canonical Monte Carlo algorithms are generalizations of random and sequential updating algorithms for lattice systems to continuum fluid models. The elementary steps, insertions and removals, are constructed by generating points in space, either at random (random updating) or in a definite order (sequential updating). The type of move is deduced by examining the local environment in the neighborhood of the selected point. Simulation results indicate that these algorithms converge faster than standard grand canonical algorithms. The efficiency enhancement of spatial updating may be understood by the following physical reasoning: successful insertion (or removal) of a particle triggers a higher probability of a successful insertion (or removal) of another particle at a neighboring point in space. These neighbor type of effects are rarely observed in standard grand canonical algorithms.

Due to the nature of the updating and the absence of strict detailed balance, spatial updating algorithms based on sequential updates are ideal for parallel implementation via domain or geometric decomposition techniques. Spatial updating algorithms are only applicable in the grand canonical ensemble, and they are distinctly different from sequential updating of particles that is commonly employed in canonical Monte Carlo simulations. Due to particle indistinguishability, sequential updating of particles in canonical Monte Carlo simulations is identical to random updating. However, sequential updating of particles can be used to reduce inter-processor communication time in parallel canonical simulations. Simulation results on Lennard-Jones systems indicate a nearly perfect improvement in parallel efficiency for large systems both for grand canonical and canonical simulations.

Simulations of phase transitions are hampered by long relaxation times associated with the unbounded growth of correlations and fluctuations in the vicinity of the critical region. Although several attempts have been made, there is no universally accepted grand canonical Monte Carlo algorithm capable of reducing the critical slowing down effects associated with phase transitions. In this work, it is shown that a combination of spatial updating with tempering techniques can increase the simulation efficiency by several orders of magnitude. In simulated tempering, several macrostates are combined to form a super-ensemble or an expanded ensemble. Each macrostate comprises a grand canonical system at a given value of the chemical potential and temperature. The elementary steps consist of particle transfers as well as transitions between different macrostates. Particle transfers are implemented according to spatial updating and transitions between different macrostates according to a heat-bath algorithm. The combination of spatial updating with the heat-bath algorithm yields an efficiency enhancement which is 2-4 orders of magnitude higher compared to standard algorithms. The crucial parameters that determine the efficiency gain are the frequency of transitions and the number of different macrostates.

The previous algorithmic advances allow for much bigger system sizes to be simulated and have thus profound ramifications in data analysis via finite-size scaling techniques. These techniques have been applied in the determination of the fluid-fluid and fluid-solid phase diagrams of systems of competing interactions, a case that is relevant in globular protein crystallization. Due to the nature of the updating, spatial updating is efficient even at high, liquid-like densities and can lead to direct determination of fluid-solid coexistence. In addition, the proposed combination of spatial updating with simulated tempering has been used to study metastable fluid-fluid separation via simulations of constrained systems. The increased efficiency of the present methodology also allows it to be used in numerical studies of homogeneous and heterogeneous nucleation to obtain: (i) nucleation barriers, (ii) free energies of cluster formation, and (iii) the critical cluster size beyond which clusters (on the average) grow.

See more of this Session: Recent Advances in Molecular Simulation Methods I
See more of this Group/Topical: Computational Molecular Science and Engineering Forum

Multiscale Optimization of Coarse-Grained Models Via the Relative Entropy

Aviel Chaimovich, Chemical Engineering, University of California Santa Barbara, Santa Barbara, CA

M. Scott Shell, Chemical Engineering, University of California Santa Barbara

Multiscale simulation methods are commonly used to study the properties of complex systems with multiple length and time scales [1]. These methods often connect detailed all-atom (AA) molecular models with simplified coarse-grained (CG) ones that enable longer and larger simulation studies. However, it has been challenging to identify optimal yet universal parameterization methodologies for the development of CG models. Recently, we introduced a rigorous theoretical framework for such a task: given an AA model, an optimal CG one can be determined by minimizing an informatic property termed the relative entropy [2]. In physical terms, the relative entropy measures deviations in the configurational ensemble probabilities as molecular features are coarse-grained away. In a previous work, it was shown that the relative entropy method naturally reproduces important inverse problems in statistical mechanics, including uniqueness theorem and variational mean field theory [2].

In this work, we present two methods for numerical optimization of CG models using relative entropy minimization. One is based on the original formulation of the relative entropy, and it involves computing property averages via simulations in both the AA and CG ensembles. Using standard free energy perturbation techniques, we present an alternative approach that requires only averages in the AA ensemble, allowing for a very fast implementation; furthermore, this reformulation of the relative entropy permits a basic expansion which yields analytical expressions for optimization in special cases. We first compare the statistical efficiency and convergence behavior of these approaches via an elementary test case, the "coarse-graining" of the nearest-neighbor lattice gas into the comparable mean-field system. We demonstrate that our mean-field solution performs in a superior manner with respect to the traditional mean-field approach. Moreover, we show that the relative entropy systematically signals the performance of the mean-field systems, thus serving as an apparent universal metric for the aptitude of CG models. Subsequently, we use the relative entropy approach to parameterize a simple (off-lattice) spherically-symmetric model of liquid water [3]. We evaluate the ability of the optimized CG water in reproducing water's bulk thermophysical properties and the hydrophobic effect [3]. Again, we show that the relative entropy serves as an indicator of the capability of these optimized models.

1. Voth, G.A., Introduction: Coarse-Graining in Molecular Modeling and Simulation. Journal of Chemical Theory and Computation, 2006. 2(3): p. 463-463.

2. Shell, M.S., The relative entropy is fundamental to multiscale and inverse thermodynamic problems. The Journal of Chemical Physics, 2008. 129(14): p. 144108-7.

3. Chaimovich, A. and M.S. Shell, Anomalous waterlike behavior in spherically-symmetric water models optimized with the relative entropy. Physical Chemistry Chemical Physics, 2009. 11(12): p. 1901-1915.

See more of this Session: Recent Advances in Molecular Simulation Methods I
See more of this Group/Topical: Computational Molecular Science and Engineering Forum

Multi-Particle Sampling in Monte Carlo Simulations On Fluids: Efficiency and Extended Implementations

Ivo Nezbeda, Chemistry Department, J. E. Purkinje University, Usti n. L., Czech Republic

Filip Moucka, Physics Department, J. E. Purkinje University, Usti n. L., Czech Republic

Various technical aspects affecting efficiency of a recently proposed novel Monte Carlo simulation scheme based on biased simultaneous displacements/rotations of all particles of the system are investigated using two polarizable models of water, the Chialvo-Cummings and Brodholt-Sampoli-Vallauri models, as a test case. Necessary expressions for polarizable site-site interaction models are derived along with a novel smoothing of the potential at the cutoff distance. In addition to the common thermodynamic and structural properties, the mean squared displacements, rotation relaxation, speed of equilibration (translational order parameter), and autocorrelation oefficients have been computed as well in order to assess efficiency of the method. Gain in speed by parallelization has also been examined.

Performance of the method is compared with both the standard one-particle move method and available approximate methods. It is shown that the multi-particle move method erforms about by a factor of 10 faster for the systems considered when compared with the common Monte Carlo scheme, and several times faster when compared with approximate methods. Parallelized codes of the multi- particle move method may then perform about seventy times faster than the conventional Monte Carlo. These conclusions hold true for the system size simulated (N=256) because the efficiency of the multi-particle method depends on the size of the system: its efficiency even increases with increasing number of particles.

See more of this Session: Recent Advances in Molecular Simulation Methods I
See more of this Group/Topical: Computational Molecular Science and Engineering Forum

A Benchmark Set for Validating and Testing Free Energy Estimation Methods in Molecular Design

Michael R. Shirts, Chemical Engineering, University of Virginia, Charlottesville, VA

There has been both intense interest and significant confusion in the simulation and theory community in determining the most efficient and reliable methods to perform free energy calculations. In this talk, I present a benchmark set for calculating free energies of molecular transformations in solution, and use this set to compare the performance of a range of published free energy methods, including thermodynamic integration, free energy perturbation, the weighted histogram analysis method, the Bennett acceptance method, transition matrix approaches, and the Wang-Landau based methods. I will describe efforts to generalize this benchmark set, including the distribution of initialization files for a number of different computational computation platforms to make it a truly general benchmark, which will hopefully provide significant guidance to the simulation field.

See more of this Session: Recent Advances in Molecular Simulation Methods I
See more of this Group/Topical: Computational Molecular Science and Engineering Forum

A New Method to Obtain Chemical Potential Differences in Periodic Molecular Simulations

Stan G. Moore, Chemical Engineering, Brigham Young University, Provo, UT

Dean R. Wheeler, Chemical Engineering, Brigham Young University, Provo, UT

The chemical potential is a very useful and important property. It can be related to many different phenomena such as phase equilibria, including solubility and osmosis; transport processes such as diffusion; as well as chemical reaction rates. These phenomena are important in many industrial applications. A new method of simulating differences in chemical potentials using molecular simulations has been developed for single-component and multi-component systems. This method uses a periodic external force-field to cause a density or concentration gradient across the simulation cell. The difference in the chemical potential across the system can then be directly related to the applied field. In contrast to methods such as Widom or Gibbs Ensemble Monte Carlo, no difficult particle insertions are necessary, so the method is well-suited for large, structured molecules and for dense liquids. We will report sample calculations for a few systems of interest, including a binary Lennard-Jones mixture and a liquid electrolyte.

See more of this Session: Recent Advances in Molecular Simulation Methods I
See more of this Group/Topical: Computational Molecular Science and Engineering Forum

GROW: a GRadient-Based Optimization Workflow for Model Building Beyond Manual Intervention

Marco Hülsmann, Computational Chemical Engineering Group, Fraunhofer Institute for Algorithms and Scientific Computing (SCAI), Sankt Augustin, Germany

Jadran Vrabec, Institute of Thermodynamics and Thermal Process Engineering, University of Stuttgart, Stuttgart, Germany

Astrid Maaß, Computational Chemical Engineering Group, Fraunhofer Institute for Algorithms and Scientific Computing (SCAI), Sankt Augustin, Germany

Thorsten Köddermann, Computational Chemical Engineering Group, Fraunhofer Institute for Algorithms and Scientific Computing (SCAI), Sankt Augustin, Germany

Dirk Reith, Computational Chemical Engineering Group, Fraunhofer Institute for Algorithms and Scientific Computing (SCAI), Sankt Augustin, Germany

Nowadays, molecular simulations are widely used to support the development process for new materials. Through simulations researchers are capable of predicting general trends quite well. But the key to quantitatively correct property predictions is the accuracy of a simulation's foundation, the force field. A force field describes the intra- and intermolecular interactions by a semi-empirical equation and its associated parameters. While the equation's functional form is usually clear, the force field parameterization is often tedious. Thereby manual adjustment and optimization is, at best, extremely time consuming. Hence, an automated parameterization scheme is essential in our pursuit to create tailor-made models for specific investigations in a timely fashion. This has been realized and implemented into a Gradient-based Optimization Workflow (GROW) for the automated development of molecular models.

GROW is a program tool kit to facilitate the use of gradient-based numerical optimization of force field parameters. Its components include various optimization algorithms (including Quasi Newton algorithms or Trust Region methods), analysis scripts, and I/O-handling. GROW can be "attached" to various standard MD simulation engines, making it a powerful companion for many application fields.

In this presentation, we will show how various algorithms perform and discuss their advantages and disadvantages. Experimental target values, used for verification purposes only, can be typically matched by GROW within a few percent for common properties (e.g. density, thermal and electrical conductivity, diffusion, and vapor pressure). This benefit will be illustrated by some atomistic force field models for small solvent molecules, polymer precursors, and ionic liquids.

See more of this Session: Recent Advances in Molecular Simulation Methods II
See more of this Group/Topical: Computational Molecular Science and Engineering Forum

Accurate Kirkwood-Buff Integrals From Molecular Dynamics Simulations

Rasmus Wedberg, Department of Chemical and Biochemical Engineering, DTU, 2800 Kgs. Lyngby, Denmark

John P. O'Connell, Chemical Engineering, University of Virginia, Charlottesville, VA

Günther Peters, Department of Chemistry, Technical University of Denmark, Kgs. Lyngby, Denmark

Jens Abildskov, Chemical and Biochemical Engineering, Technical University of Denmark, Kgs. Lyngby, Denmark

We seek to develop a rigorous approach for extracting Kirkwood-Buff (KB) integrals, i.e. spatial integrals of the molecular pair-distribution functions, from molecular dynamics simulations in the NVT or NPT ensemble. Computing the KB integrals of a mixture at different compositions allows one to calculate several derivatives of thermodynamic functions (activities, partial molar volumes, isothermal compressibility) easily.

However, direct numerical integration of the g(r) obtained from MD is fragile since the tail deviates from the desirable, grand-canonical distribution function. Various empirical approaches to extend the MD results to large r, and thus obtain an accurate integral, have been explored previously. In this work, we utilize the asymptotic behavior of the direct correlation function, c(r), which at large r approaches the Mayer function, and connect this with the total correlation function, h(r), via the Ornstein-Zernike relation.

The g(r) obtained from MD is assumed to be correct for $r < R$ for a certain R. An extension is sought, that is consistent with a c(r) exhibiting the proper behavior when $r > R$. Together with the requirement that g(r) and c(r) satisfy the Ornstein-Zernike equation one obtains a closed form integral equation, which is efficiently solved using Newton's method. This resembles an approach proposed by Verlet [L. Verlet, 1968. Phys. Rev., 165(1): 201].

The method is tested on the Lennard-Jones fluid. MD simulations are carried out covering a broad range of temperatures and densities. Verification of the results is done by comparing the isothermal compressibilities derived from the KB integrals with those obtained from the equation of state by Johnson et. al. [J.K. Johnson, J.A. Zollweg, K.E. Gubbins, 1993. Mol. Phys. 78(3): 591], which results from smoothing PVT data obtained from a large number of MD simulation with the Lennard-Jones potential. The two routes are found to be in good agreement.

The paper describes the results obtained so far, and results of current work on rationalizing the choice of R, and extending the method to angle-dependent potentials.

See more of this Session: Recent Advances in Molecular Simulation Methods II
See more of this Group/Topical: Computational Molecular Science and Engineering Forum

A Simple Simulation Method for Calculating the Free Energy of Atomic Solid

Sang Kyu Kwak, Division of Chemical and Biomolecular Engineering, Nanyang Technological University, Singapore, Singapore

Prediction for the free energy of solid has been considered as a difficult task even though there exist several methods, of which implementations are generally non-trivial and even more if one expects such a high accuracy. We introduce a simple simulation method to predict the free volume of particle in crystal, consequently configurational entropy. The idea of the generic Monte Carlo integration has been reversely used in this approach. In prediction of excess Helmholtz free energies to confirm the correctness of the method, we chose close-packed hard sphere crystals - fcc and hcp. Differences in calculated free energies are also provided to indicate the accuracy of the method. This robust scheme works for wide densities over melting to closest-packed without cumbersome parameter adjustments and for other potential models such as Lennard-Jones and square-well with relatively simple modifications.

See more of this Session: Recent Advances in Molecular Simulation Methods II
See more of this Group/Topical: Computational Molecular Science and Engineering Forum

Limitations and Recommendations for the Calculation of Shear Viscosity Using Reverse Nonequilibrium Molecular Dynamics

Craig M. Tenney, Chemical and Biomolecular Engineering, University of Notre Dame, Notre Dame, IN

Edward J. Maginn, Chemical and Biomolecular Engineering, University of Notre Dame, Notre Dame, IN

The reverse nonequilibrium molecular dynamics (RNEMD) method (Muller-Plathe, Phys Rev 59, 1999) calculates the shear viscosity of a fluid by imposing a non-physical exchange of momentum and measuring the resulting shear velocity gradient. In contrast with most other nonequilibrium methods, because linear momentum and energy can be conserved in RNEMD, coupling to an external thermostat is not strictly required. RNEMD has been used to calculate the viscosities of fluids with complexities varying from simple atoms to complex molecules. It is known that RNEMD will fail to yield linear shear velocity profiles at high momentum flux rates. In this study we investigate the range of momentum flux values over which RNEMD yields usable (linear) velocity gradients. We find that non-linear velocity profiles result primarily from gradients in fluid temperature and density. The temperature gradient results from conversion of heat into bulk kinetic energy, which is transformed back into heat elsewhere via viscous heating. An expression is derived to predict the temperature profile resulting from a specified momentum flux for a given fluid and simulation cell. Although primarily bounded above, we also describe milder low-flux limitations. RNEMD results for a Lennard-Jones fluid agree with equilibrium molecular dynamics and conventional nonequilibrium molecular dynamics calculations at low shear, but RNEMD under-predicts viscosity relative to conventional NEMD at high shear.

See more of this Session: Recent Advances in Molecular Simulation Methods II
See more of this Group/Topical: Computational Molecular Science and Engineering Forum

Study of Proton Transport Using Reactive Molecular Dynamics

Myvizhi Esai Selvan, Department of Chemical and Biomolecular Engineering, University of Tennessee, Knoxville, TN

David J. Keffer, Department of Chemical and Biomolecular Engineering, University of Tennessee, Knoxville, TN

Shengting Cui, Department of Chemical and Biomolecular Engineering, University of Tennessee, Knoxville, TN

Stephen J. Paddison, Department of Chemical and Biomolecular Engineering, University of Tennessee, Knoxville, TN

The proton transport occurs through a combination of vehicular diffusion - movement of the center of mass of hydronium ions and structural diffusion - transfer of protons among water molecules. We have developed a coarse-grained reactive molecular dynamics (RMD) algorithm to model the structural diffusion of proton using the classical molecular dynamics simulation. The major advantage of this approach is that we use computationally efficient, well established non-reactive potentials that already exists and has been optimized. The algorithm implements the structural diffusion via three steps (i) satisfaction of the triggers, (ii) instantaneous reaction and (iii) local equilibration. A set of triggers were chosen based on the transition state obtained from *ab initio* calculation and activation energy to check for the favorable configuration for the structural diffusion of proton to take place. These trigger values were parameterized to fit the experimental values of the rate constant in order to capture the essential molecular and macroscopic features of structural diffusion. The final step is to satisfy the heat of reaction and ensure a correct ending configuration after the proton hopping.

We validated this algorithm by modeling the structural diffusion of proton in extensively studied system like bulk water. A single hydrated proton was studied in a system of 650 water molecules as a function of temperature from 280 K to 320 K. The triggers were chosen based on the structure of the dominant hydrated models (Zundel and Eigen cations) necessary for the structural diffusion of protons and were parameterized to reproduce the experimental rate constant and activation energy. Using this procedure, an in depth study of the correlation and contribution of the two components of the diffusion to the charge diffusion was conducted. The algorithm is generalized by the environment sensitive geometric and energetic trigger. The RMD endeavors to have a valid starting and ending configuration for the structural diffusion. Information regarding the transition state is embedded in the triggers, but the algorithm makes no attempt to dynamically describe the structure of the transition state. Therefore, functional extrapolation of structural diffusion in one environment (bulk water) to another environment (hydrated proton exchange membrane (PEM)) is enabled by the additional degree of coarse graining. Recognition and adaptability of the environment by the algorithm can be proved by studying proton transport in hydrochloric acid at different concentrations with the trigger values parameterized for infinite dilution.

The next step is to model proton transport in hydrated PEM. The structural diffusion of proton in hydrated membrane can be explained by the following three reactions.

$$H_3O^+ + H_2O \xleftrightarrow{4H_2O} H_2O + H_3O^+ \quad (1)$$

$$H_3O^+ + H_2O \xleftrightarrow{nH_2O/(4-n)SO_3^-} H_2O + H_3O^+ \quad (2)$$

$$SO_3H + H_2O \xleftrightarrow{nH_2O} SO_3^- + H_3O^+ \quad (3)$$

Equation 1 is similar to proton transport in bulk water and will take place along the center of the water channels and equation 2 and 3 will occur at the interface of the hydrophobic and hydrophilic regions. RMD algorithm can accommodate all the above reactions and the triggers can register the presence of the sulfonic acid groups and other environmental factors like confinement. Thus the algorithm will allow us to measure the diffusivities in the hydrated membranes and provide a molecular level understanding of how features of the membrane such as enhanced confinement and hydration impact proton mobility.

Acknowledgements

The work is supported by a grant from the U. S. Department of Energy BES under the contract number DE-FG02-05ER15723. This research used resources of the Center for Computational Sciences at Oak Ridge National Laboratory, which is supported by the Office of Science of the DOE under Contract DE-AC05-00OR22725. This work also used resources of the National Institute for Computational Sciences (NICS), supported by NSF with agreement number: OCI 07-11134.

See more of this Session: Recent Advances in Molecular Simulation Methods II
See more of this Group/Topical: Computational Molecular Science and Engineering Forum

Field-Biased Molecular Simulation Methods for Hard Sphere Polymer Chains

J. Richard Elliott, Chemical and Biomolecular Engineering, University of Akron, Akron, OH

Amir Vahid, Chemical and Biomolecular Engineering, University of Akron, Akron, OH

External fields can be used to impose density profiles in inhomogeneous fluids and interfacial phenomena1.[1] In this study an electric field has been imposed on 1372 hard spheres through 20 negative point charges and 20 positive charges. Also, the effect of partial charges was investigated on a polyelectrolyte with implicit and explicit solvent.[2] Long-range interactions are considered through particle-mesh Ewald summation and its pairwise alternatives.[3] It has been found that it is not necessary to update the Coulombic interactions after each time-step. Energy is conserved even after many numbers of time-steps. Therefore, the computation time for the long-range interaction is less than the discontinuous molecular dynamic (DMD) and/or discontinuous Monte Carlo components. This means that the forced-biased discontinuous molecular simulation method is viable for future studies of confined fluids containing interface and ionic liquids as performed with a field-biased conventional molecular dynamic method by Wardle et al.[4, 5]

(1) Hansen, J.-P.; McDonald, I. R., Theory of Simple Liquids. 3 rd ed.; Academic Press: Amsterdam, 2006.

(2) Reddy, G.; Yethiraj, A. Macromolecules 2006, 39, (24), 8536-8542.

(3) Fennell, C. J.; Gezelter, J. D. Journal of Chemical Physics 2006, 124, (23).

(4) Wardle, K. E.; Carlson, E.; Henderson, D.; Rowley, R. L. Journal of Chemical Physics 2004, 120, (16), 7681-7688.

(5) Wardle, K. E.; Henderson, D. J.; Rowley, R. L. Fluid Phase Equilibria 2005, 233, (1), 96-102.

See more of this Session: Recent Advances in Molecular Simulation Methods II
See more of this Group/Topical: Computational Molecular Science and Engineering Forum

Field Theoretic Simulations in the Gibbs Ensemble

Robert A. Riggleman, Department of Chemical Engineering, University of California, Santa Barbara, Santa Barbara, CA

Glenn H Fredrickson, Department of Chemical Engineering and Materials Department, University of California, Santa Barbara, Santa Barbara, CA

Performing field theoretic simulations of phases in equilibrium is a computationally demanding problem. To obtain properties of both phases from one simulation (such as the concentration of a species in each phase), it is necessary to perform simulations large enough so that the interface between the two phases does not affect the estimate of the bulk properties. In this talk, I will describe a new method we have developed for performing field theoretic simulations in the so-called Gibbs ensemble, where two simulation boxes are kept in chemical and mechanical equilibrium by allowing the boxes to exchange both particles and volume. By maintaining a constant total number of particles and total volume, such a simulation can efficiently simulate two bulk phases in equilibrium in the canonical ensemble, allowing a reliable estimate of the properties of the two phases. Our method will be demonstrated in both the mean-field limit and in a simulation that fully samples the fluctuations of the field theory.

See more of this Session: Recent Advances in Molecular Simulation Methods II
See more of this Group/Topical: Computational Molecular Science and Engineering Forum

High-Throughput Catalysis and Information Management

Michael Doyle, Accelrys Inc., San Diego, CA

Experimental teams developing catalysts produce hundreds of samples per year. Without a proper information management system, information on how they were prepared and how they performed can become inaccessible, or at worst, completely missing. This results in increased cost and lower product performance. A data management tool together with predictive analytics can increase R&D efficiency and reduce the number of experiments that need to be performed. In this presentation we explore some of the tools available such as recursive partitioning, principal component analysis, and neural networks; and demonstrate how these techniques have been used in catalyst discovery. Enriching the available data with the results of molecular modeling provides valuable information that can improve the reliability of predictive models.

See more of this Session: Recent Advances in Molecular Simulation Methods II
See more of this Group/Topical: Computational Molecular Science and Engineering Forum

Modeling Electronic and Steric Effects On An Immobilized (salen)Mn Catalyst

Gloria A. E. Oxford, Chemical and Biological Engineering, Northwestern University, Evanston, IL

Randall Q. Snurr, Chemical and Biological Engineering, Northwestern University, Evanston, IL

Linda J. Broadbelt, Chemical and Biological Engineering, Northwestern University, Evanston, IL

Metal-organic frameworks (MOFs) show great promise for combining the best features of homogeneous and heterogeneous catalysis. MOFs are nanoporous materials formed from self-assembly of metal ion vertices and organic linkers into regular crystalline structures. An appealing strategy for developing MOF catalysts is to use a homogeneous asymmetric catalyst as the organic linker [1], but immobilization of the catalysts in the crystal presents a challenge for preserving high enantioselectivity. Recently, a MOF was synthesized from a (salen)Mn catalyst and biphenyldicarboxylate as linkers with zinc paddlewheel corners [1]. The MOF performed enantioselective epoxidation catalysis with ee values only slightly lower than the homogeneous catalyst (82% ee vs. 88% ee). It is unknown whether electronic effects or steric effects of the MOF environment lead to the decrease in enantioselectivity. In this work, a combination of modeling techniques is used to decouple these effects and determine the dominating factor affecting the asymmetric induction of the catalyst in the MOF.

It is well known that the electronic nature of the 5,5'-substituents on the salen ligand strongly affects the enantioselectivity of the catalyst [2], and density functional theory (DFT) calculations have correlated the modified Hammett parameter σ^+ of the 5,5'-substituents to reactivity properties [3]. To study the electronic effect of coordination of the 5,5'-substituents to zinc ions in the MOF structure on the enantioselectivity of the (salen)Mn catalyst, hybrid quantum mechanics/molecular mechanics (QM/MM) calculations were performed first to validate these methods for investigating electronic effects of salen ligands and second to determine the electronic effects of the ligand coordination in the MOF. From these calculations, it can be shown whether the coordination geometry in the MOF has a significant effect on the catalysis.

The direction in which the reactant approaches the active site is also believed to play an important role in the asymmetric induction of the (salen)Mn catalyst [4]. To understand the role of the MOF in guiding the reactant to the active site, classical optimizations and molecular dynamics simulations have been performed. These calculations reveal the preferred approach to the homogeneous catalyst and to the catalyst in the MOF. They also provide information about the preferred location of the reactant molecules within in the pores of the MOF. With the insight gained from these calculations as well as from the QM/MM calculations, more effective (salen)Mn MOF catalysts can be designed.

References

[1] S.-H. Cho, B. Ma, S. T. Nguyen, J. T. Hupp and T. E. Albrecht-Schmitt, Chem. Commun., 2563 (2006).

[2] E. N. Jacobsen, W. Zhang and M. L. Guler, J. Am. Chem. Soc. 113, 6703 (1991).

[3] L. Cavallo and H. Jacobsen, J. Org. Chem. 68, 6202 (2003).

[4] T. Katsuki, Adv. Synth. Catal. 344, 131 (2002).

See more of this Session: Poster Session: Computational Molecular Science and Engineering Forum
See more of this Group/Topical: Computational Molecular Science and Engineering Forum

Environmental Effects On Catalytic Activity at the RuO2(110) Surface

Hangyao Wang, Chemical and Biomolecular Engineering, University of Notre Dame, Notre Dame, IN

William F. Schneider, Department of Chemical and Biomolecular Engineering, Department of Chemistry and Biochemistry, University of Notre Dame, Notre Dame, IN

A key challenge in atomistic simulation of heterogeneous catalysis is to capture the effects of reaction environment on surface reactivity. The energetics, mechanisms, and even chemical species present at a reacting surface can be a strong function of the conditions to which the surface is exposed. In this work, we use DFT calculations and thermodynamic and kinetic modeling to examine two examples of such effects at the RuO2(110) surface. We first consider O2 adsorption and dissociation, and show that the molecular O2 species observed in TPD experiments and identified as a precursor to O2 dissociation is in fact a spectator present only at high coverages of surface O. The existence of "stranded" adsorption sites along the one-dimensional surface rows is shown to produce these stranded sites. We next consider catalytic CO oxidation and its competition with surface poisoning. We identify both carbonate and bicarbonate surface poisons and show that the coverage of the latter is highly sensitive to water concentration and likely accounts for the surface poisoning observed experimentally.

See more of this Session: Poster Session: Computational Molecular Science and Engineering Forum
See more of this Group/Topical: Computational Molecular Science and Engineering Forum

Ion Adsorption and Electrode-Electrolyte Interfacial Structure Effects On Oxygen Adsorption

Kuan-Yu Yeh, Chemical Engineering, Pennsylvania State University, University Park, PA

Michael J. Janik, Chemical Engineering, Pennsylvania State University, University Park, PA

The accurate representation of electrode potential variation, interfacial electric fields, and interfacial solvation is challenging within density functional theory (DFT) models. Accurate evaluation of elementary electrocatalytic reaction step energetics requires models to evaluate electrode and electrolyte effects and the dependence of kinetics on electrode potential. Several approaches have been developed to model the electrical double layer using first-principles methods. The first model is a typical vacuum slab calculation with the linear dependence of electron energy on potential. An external, homogeneous electric field model provides a computationally inexpensive model of an electrified interface, but correlation to a specific electrode potential is unclear. A third model directly controls the electrode potential by charging the aqueous electrochemical system. These models will be compared in their evaluation of the impact of electrode potential variation on molecular oxygen adsorption to the solvated Pt(111) surface, of relevance to the oxygen reduction reaction within a proton exchange membrane fuel cell (PEMFC).

The interfacial water structure affects the binding strength and extent of charge transfer for molecular oxygen adsorbed on the Pt(111) surfaces. In a PEMFC, the electrolyte consists of hydrated Nafion ionomers. The presence of other adsorbing species may alter electronic interactions and affect the adsorbed oxygen state. We use solvated sulfate groups as our electrolyte, and investigate the influence of ion adsorption on oxygen adsorption. Three electrochemical models will be use to differentiate the importance of local electric field, through surface and through space interactions, coadsorption, and solvent effects, especially ion adsorption.

See more of this Session: Poster Session: Computational Molecular Science and Engineering Forum
See more of this Group/Topical: Computational Molecular Science and Engineering Forum

Cluster-Expansion-Based Modeling of the Coverage Dependence of Adsorbate Binding at a Metal Surface

David J. Schmidt, Chemical and Biomolecular Engineering, University of Notre Dame, Notre Dame, IN

William F. Schneider, Chemical and Biomolecular Engineering, University of Notre Dame, Notre Dame, IN

Wei Chen, McC Materials Science and Engineering, Northwestern University, Evanston, IL

Christopher M. Wolverton, McC Materials Science and Engineering, Northwestern University, Evanston, IL

The binding of adsorbates to surfaces is often characterized using mean field models. This approach treats adsorbate binding energies as constant independent of coverage and neglects and correlations between the locations of adsorbates. In this work, we use lattice-based cluster expansions to model the configuration-dependent energetics of oxygen adsorption on the Pt(111) surface. The cluster expansions were parameterized using DFT calculations of more than 30 distinct configurations, the convergence of the expansion with cluster and configuration size evaluated, and the relative contributions of electronic and relaxation effects to adsorbate interactions evaluated. Surface relaxations were found to have an appreciable but not dominate effect on the formation energies. The cluster expansion was used in concert with Monte Carlo simulations to model adsorbate binding energy (chemical potential) as a function of surface coverage and temperature. The binding energy was found to be roughly linear with respect to the surface coverage, consistent with earlier studies.

See more of this Session: Poster Session: Computational Molecular Science and Engineering Forum
See more of this Group/Topical: Computational Molecular Science and Engineering Forum

Step Formation On Yttria-Stabilized Zirconia: Combining DFT and Reactive Force Fields to Interpret Surface Characterization

Adam D. Mayernick, Department of Chemical Engineering, Pennsylvania State University, University Park, PA

Michael J. Janik, Chemical Engineering, Pennsylvania State University, University Park, PA

Adri C.T. van Duin, Department of Mechanical and Nuclear Engineering, Pennsylvania State University, University Park, PA

Matthias Batzill, Department of Physics, University of South Florida, Tampa, FL

Yttria-stabilized zirconia (YSZ) is a mixed oxide with high oxygen ion conductivity due to the presence of Y^{3+} induced oxygen defects within the cubic phase of zirconia. The high oxygen conductivity of YSZ facilitates its application in a variety of applications such as oxygen sensors, solid oxide fuel cell (SOFC) membranes, and heterogeneous catalyst supports. The surface structure of YSZ influences its performance in each of these applications, motivating characterization of stable YSZ surfaces at the atomistic scale. We report the results of density functional theory (DFT) calculations as well as reactive force field (ReaxFF) molecular dynamics and Monte Carlo simulations used to interpret and supplement scanning tunneling microscopy (STM) imaging of the YSZ(111) surface.

Imaging of the YSZ(111) surface reveals a high concentration of monatomic height step edges, formed upon Ar-ion sputtering and vacuum annealing. We calculate surface formation energies within the Vienna Ab-initio Simulation Program (VASP) for pure ZrO_2 and YSZ, for both flat and stepped surfaces. The surface formation energy of the stepped YSZ(111) surface is less than that of the pure ZrO2 stepped surface, showing that the addition of Y to cubic zirconia stabilizes monatomic steps on the (111) termination. To allow consideration of larger scale step structures, ReaxFF was used to perform molecular dynamics to simulate annealing. Surface energies calculated by DFT are used to parameterize the ReaxFF force field to facilitate accurate treatment of the YSZ surface. We apply Monte Carlo methods to sample the possible sites for yttrium and oxygen vacancies on the stepped YSZ(111) surface to find a minimum energy configuration. We perform simulated annealing on our minimum energy stepped surface model to probe surface reorganization and step stability. Our results show that Yttria-doping of cubic zirconia stabilizes the formation of monatomic steps on the (111) surface, in agreement with STM results.

See more of this Session: Poster Session: Computational Molecular Science and Engineering Forum
See more of this Group/Topical: Computational Molecular Science and Engineering Forum

Water Adsorption Behavior On Unique Titanium Dioxide Polymorph (TiO2-B) Facets

Xiaohua Lu, State Key Laboratory of Materials-oriented Chemical Engineering, Nanjing University of Technology, Nanjing, China

Vega Lew, State Key Laboratory of Materials-oriented Chemical Engineering, Nanjing University of Technology, Nanjing, China

Xiaojing Guo, State Key Laboratory of Materials-oriented Chemical Engineering, Nanjing University of Technology, Nanjing, China

Linghong Lu, State Key Laboratory of Materials-oriented Chemical Engineering, Nanjing University of Technology, Nanjing, China

Titanium dioxide (TiO2) processes the promising perspective in many applications, such as solar cell, photocatalysis, catalyst support, water splitting devices and sensors etc. TiO2-based materials received enormous expectation in solving or easing the environmental challenges and energy crisis. Surface chemical and physical properties play a dominate role in most fields, such as biology, materials science, catalysis etc. It is well accepted that surfaces with high reactivities are extraordinary important both from scientific and technological viewpoint. Current investigations focus on three main crystallographic forms of TiO2, rutile, anatase and brookite. But there is another polymorph of TiO2 named TiO2-B in ambient conditions, which has been rarely studied since it was identified in 1980s. To advance the basic understanding this polymorph of TiO2, we present ab initio calculations to reveal the structure and electronic properties of TiO2-B crystal.

Water is a most common molecule in nearly all the application fields of TiO2. H2O molecules are also widely adopted to probe the surface chemistry of various metal oxides. Water adsorption or dissociation on different surfaces is one of the most important issues for understanding and controlling functionality of metal oxide materials. Moreover, we studied water adsorption behaviors on TiO2-B facets, (001) and (100), to elaborate the chemi-/physic- properties. Water distinct adsorption behaviors on two surfaces, (001) and (100) are identified. Dissociative adsorption of water is quite favored, even at monolayer coverage on TiO2-B (100). Whereas, molecular adsorption of water is the most favorable state on TiO2-B (001), regardless of the coverage. Thus, we reported that another surface of the forth TiO2 polymorph, TiO2-B (100), would be probably very reactive as anatase (001) or rutile (011)-2x1, on which dissociation of water is also favored. These results highligh surface control during the synthesis process of the catalyst, it is another key issue for improving the catalysis performance.

See more of this Session: Poster Session: Computational Molecular Science and Engineering Forum
See more of this Group/Topical: Computational Molecular Science and Engineering Forum

Structural and Dynamic Properties of Interfacial Water at the Silica Surface

Dimitrios Argyris, School of Chemical Biological and Materials Engineering, The University of Oklahoma, Norman, OK

Paul D. Ashby, Materials Sciences Division, Lawrence Berkeley National Laboratory, Berkeley, CA

Alberto Striolo, School of Chemical Biological and Materials Engineering, The University of Oklahoma, Norman, OK

The molecular structure of interfacial water at the silica solid surfaces was investigated by conducting equilibrium molecular dynamics simulations. Three crystallographically equivalent $SiO2$ substrates are considered. Different degrees of surface hydroxylation where employed to assess the effect of the surface chemistry on the structure of interfacial water. The degree of hydroxylation of the unsaturated oxygen atoms varied from 100, 50, and 0%, respectively. The SPC/E model was used to model water-water interactions. Density profiles, in-plane radial distribution functions, in-plane density distribution, hydrogen bond profiles, residence probability, reorientation correlation functions, and hydrogen bond lifetime were calculated to quantify the structural and dynamic behavior of water. Our results show that the surface hydroxylation affects the structure, orientation, hydrogen bond network, and dynamic behavior of interfacial water molecules. Data analysis suggests that the degree of hydroxylation controls the amount of water molecules in the first interfacial layer as well as the distance between the first adsorbed layer and the substrate. Well-organized and uniform structures of interfacial water appear on the homogeneously hydroxylated surface, while a heterogeneous interfacial structure, characterized by extensive water-water hydrogen bonds, forms on the partially hydroxylated surface. We demonstrate that both the local surface chemistry and water-water hydrogen bonds are the dominant factors that determine the structural properties of interfacial water. Atomic force microscopy experiments were also performed for the investigation of interfacial water on quartz. The data collected were analyzed with the Brownian force profile reconstruction technique.

See more of this Session: Poster Session: Computational Molecular Science and Engineering Forum
See more of this Group/Topical: Computational Molecular Science and Engineering Forum

The role of Pore Size and Polar Functional Groups in Hydrogen Adsorption in Metal Organic Frameworks

Nethika S. Suraweera[1] , Ruichang Xiong[1], J. P. Luna[1], Donald M. Nicholson[2] , David J. Keffer[1],

[1]Department of Chemical and Biomolecular Engineering, University of Tennessee, Knoxville, TN 37996
[2]Computer Science and Mathematics Division, Oak Ridge National Laboratory

Abstract

Efficient storage of hydrogen is an important aspect in utilizing hydrogen as a pollution free energy source. The adsorptive storage of molecular hydrogen on microporous materials plays a vital role in current studies and Metal-organic frameworks (MOFs) are one of the best hydrogen adsorbents, which exhibit high hydrogen adsorption uptakes. Isoreticular Metal-organic frameworks (IRMOF) materials which consist of zinc oxide clusters connected by organic linker molecules, have nanoscale pores and large surface area, that result in high hydrogen adsorption.

The objective of this study is to model adsorption of hydrogen in IRMOF-1, IRMOF-2, IRMOF-3, IRMOF-7, IRMOF-8, IRMOF-10, variants of IRMOF-10 with amine groups at two positions and Bromine at two positions by Path Integral Grand Canonical Monte Carlo (PI-GCMC) simulations using standard force fields. From the simulations, hydrogen adsorption isotherms at 300K and 77 K are generated for low pressure conditions (up to 10 atm). Isotherms generated are in fairly good agreement with existing experimental data and other simulation data. We characterize the effect of pore size and the presence of polar functional groups (amines and Bromine) on the adsorption isotherms.

Keywords: hydrogen adsorption, PI-GCMC simulation, adsorption isotherm, MOF, IRMOF

A Computational Study of the Behavior of Ionic Liquids Confined Inside Nanoporous Carbons

Joshua D. Monk, Cain Department of Chemical Engineering, Louisiana State University, Baton Rouge, LA

Ramesh Singh, Cain Department of Chemical Engineering, Louisiana State University, Baton Rouge, LA

Francisco R. Hung, Cain Department of Chemical Engineering, Louisiana State University, Baton Rouge, LA

Research of ionic liquids confined inside nanoporous electrodes has gained interest for their potential applications in electrochemical double layer capacitors (EDLCs) for hybrid vehicles, and dye-sensitized solar cells (DSSCs). A fundamental understanding of the behavior of ionic liquids when confined inside nm-sized pores is necessary to optimize these applications, as the properties of fluids inside nanopores can be substantially different from those exhibited by bulk phases. Using molecular dynamics simulations, we investigate the physical behavior of a typical ionic liquid (e.g., [DMIM+][Cl-]) confined inside nanoporous carbons. Different model carbon materials (e.g., slit-like graphitic pores, ordered mesoporous carbons such as CMK-3) with different pore morphologies and sizes (0.9 nm to 4.5 nm) are considered in our studies. We analyze and discuss the effects of pore size and pore morphology on the structural and dynamical properties of the confined ionic liquid.

See more of this Session: Poster Session: Computational Molecular Science and Engineering Forum
See more of this Group/Topical: Computational Molecular Science and Engineering Forum

Theoretical Study On the Kinetics and Mechanisms of the Oxidation of Phenol by Advanced Oxidation Processes: Some Insights On QM Calculations as Applied to Chemical Engineering Studies

Bruno Ramos, Escola Politécnica - Chemical Engineering Department, Universidade de São Paulo, São Paulo, Brazil

Antonio Carlos S. C. Teixeira, Chemical Engineering Department, Universidade de São Paulo, São Paulo, Brazil

It has been recently proposed that Molecular Modeling (MM) would be an essential tool of research in Chemical Engineering. In fact, it has been suggested to be in the core of the emerging paradigm in Chemical Engineering. How accurate is this? What exactly, in the present time, computational chemistry calculations may offer to Chemical Engineers? In this work, some conventional MM techniques were tested to study part of the reactions involved in the degradation of phenol by Advanced Oxidation Processes (AOP), aiming: (i) the identification of molecular-reactions intermediates, (ii) the evaluation of a method to obtain ab initio kinetic constants estimations, (iii) the study of the effect of water solvation in the calculations, and, (iv) by comparison with the available experimental data, discuss different theoretical methods with respect to the accuracy and correlation with experiment.

The results suggest that DFT (Density Functional Theory) functionals play a better role in thermodynamical and kinetic parameters calculation than the semi-empirical AM1 (Austin Model 1) or the ab initio HF (Hartree-Fock) for these systems. As for the algorithms used, it was found that Eigenvector Following with tighter convergence criteria for geometry optimizations offers a higher accuracy to experiment than Steepest Descent method or a standard set of convergence criteria. For the only elementary reaction with experimental data available, it was found that the activation free energy calculated by DFT shows a deviation of 4 kcal mol-1 (the experimental uncertainty of energy measurements is reported as 2 kcal mol-1). The kinetic constant found for this reaction is within the order found in experiments. Applying the same set of functional, basis functions and mathematical convergence criteria, other reactions were studied and new, unreported and experimentally unavailable kinetic constants were found for elementary reactions within the process. Three different solvation models were tested, and the results show that the semi-empirical AM1 model, combined with SM5.42R solvation model, exhibits a very good agreement with the experimental value for phenol. However, since the reactions in this process involve electron transfer and the solvation stabilization relies strongly upon hydrogen-bonding interactions, explicit models using water molecules clusters around the solute molecule were also tested and are reported in this work.

Under the lights of the results obtained and their correlation with experimental data, it is reasonable to suggest that MM can yes be used to produce estimations of kinetic constants. It should be stressed, however, some important aspects on the choice of the method. Molecular Modeling involves a myriad of methods and models, and it is important that the researchers know how to make a choice between these countless options. The choice depends basically on five factors: (i) the system under study; (ii) the level of accuracy desired; (iii) the time available; (iv) the hardware available; (v) the parameterization of the model (if a parameterized model is to be used). The results of this work are compared with a previous theoretical study, stressing the importance of the knowledge of the chemical system being dealt with, and the physical assumptions underlying the different MM models applied to describe it.

See more of this Session: Poster Session: Computational Molecular Science and Engineering Forum
See more of this Group/Topical: Computational Molecular Science and Engineering Forum

Mechanism Development for Combustion of Morpholine

Nicole Labbe, University of Massachusetts, Amherst, MA

Phillip R. Westmoreland, Chemical Engineering, University of Massachusetts Amherst, Amherst, MA

Arnas Lucassen, Department of Chemistry, Bielefeld University, Bielefeld, Germany

Patrick Oßwald, Department of Chemistry, Bielefeld University, Bielefeld, Germany

Ulf Struckmeier, Department of Chemistry, Bielefeld University, Bielefeld, Germany

Katharina Kohse-Höinghaus, Department of Chemistry, Bielefeld University, Bielefeld, Germany

Tina Kasper, Sandia National Laboratories, Livermore, CA

Nils Hansen, Sandia National Laboratories, Livermore, CA

Terrill A. Cool, School of Applied and Engineering Physics, Cornell University, Ithaca, NY

With the emergence of biofuels, oxygen- and nitrogen-containing fuels are becoming of new interest in combustion science. The fuel additive morpholine (1-oxa-4-aza-cyclohexane) is appropriate to study relevant combustion chemistry features because in one molecule, it mixes structural features of cycloalkanes, ethers, and amines.

Our collaboration at the Advanced Light Source of Lawrence Berkeley National Laboratory has measured spatial profiles of mole fraction in a premixed flat flame under low-pressure, fuel-rich (phi=1.3) conditions [1]. We were able to identify a large number of radical and stable species in the flame. Based on mechanistic insights from experiments and modeling of cyclohexane flames [2], we proposed probable reaction paths that should occur in a mechanism for morpholine combustion.

Gaussian 03 [3], ideal-gas thermochemistry from a rigid-rotor/harmonic-oscillator representation, transition-state theory, Bimolecular Quantum-RRK reaction theory [4], and the PREMIX/CHEMKIN flat-flame codes [5,6] were used to develop a morpholine model quantitatively. The Complete Basis Set method CBS-QB3 was employed to obtain thermochemical and kinetics parameters for key species and reactions in the mechanism. For many reactions, rate constants are adapted from cyclohexane kinetics, corrected for heat of reaction and reaction path degeneracy. When compared to our experimental data provided, we find good correlation between the experimental species profiles and our predictions.

Acknowledgments. This work is supported by the Division of Chemical Sciences, Geosciences, and Biosciences, the Office of Basic Energy Sciences, the U. S. Department of Energy, in part under grants DE-FG02-91ER14192 (P.R.W.) and DE-FG02-01ER15180 (T.A.C.). Sandia is a multi-program laboratory operated by Sandia Corporation, a Lockheed Martin Company, for the National Nuclear Security Administration under contract DE-AC04-94-AL85000. The Advanced Light Source is supported by the Director, Office of Science, Office of Basic Energy Sciences, Materials Sciences Division, of the U.S. Department of Energy under Contract No. DE-AC02-05CH11231 at Lawrence Berkeley National Laboratory. Support is also provided by the Deutsche Forschungsgemeinschaft under contract Ko 1363/18-3 (A.L., P.O., U.S., K.K.-H).

[1] A. Lucassen, P. Oßwald, U. Struckmeier, K. Kohse-Höinghaus, T. Kasper, N. Hansen, T.A. Cool, P.R.Westmoreland. "Species identification in a laminar premixed low-pressure flame of morpholine as a model substance for oxygenated nitrogen-containing fuels." Proceedings of the Combustion Institute 32, 1269-1276 (2009).

[2] M.E. Law, P.R. Westmoreland, T.A. Cool, J. Wang, N. Hansen, T. Kasper. "Benzene Precursors and Formation Routes in a Stoichiometric Cyclohexane Flame." Proceedings of the Combustion Institute 31, 565-573 (2007).

[3] Gaussian 03, Revision C.02, M. J. Frisch, G. W. Trucks, H. B. Schlegel, G. E. Scuseria, M. A. Robb, J. R. Cheeseman, J. A. Montgomery, Jr., T. Vreven, K. N. Kudin, J. C. Burant, J. M. Millam, S. S. Iyengar, J. Tomasi, V. Barone, B. Mennucci, M. Cossi, G. Scalmani, N. Rega, G. A. Petersson, H. Nakatsuji, M. Hada, M. Ehara, K. Toyota, R. Fukuda, J. Hasegawa, M. Ishida, T. Nakajima, Y. Honda, O. Kitao, H. Nakai, M. Klene, X. Li, J. E. Knox, H. P. Hratchian, J. B. Cross, V. Bakken, C. Adamo, J. Jaramillo, R. Gomperts, R. E. Stratmann, O. Yazyev, A. J. Austin, R. Cammi, C. Pomelli, J. W. Ochterski, P. Y. Ayala, K. Morokuma, G. A. Voth, P. Salvador, J. J. Dannenberg, V. G. Zakrzewski, S. Dapprich, A. D. Daniels, M. C. Strain, O. Farkas, D. K. Malick, A. D. Rabuck, K. Raghavachari, J. B. Foresman, J. V. Ortiz, Q. Cui, A. G. Baboul, S. Clifford, J. Cioslowski, B. B. Stefanov, G. Liu, A. Liashenko, P. Piskorz, I. Komaromi, R. L. Martin, D. J. Fox, T. Keith, M. A. Al-Laham, C. Y. Peng, A. Nanayakkara, M. Challacombe, P. M. W. Gill, B. Johnson, W. Chen, M. W. Wong, C. Gonzalez, and J. A. Pople, Gaussian, Inc., Wallingford CT, 2004.

[4] P.R. Westmoreland, J.B. Howard, J.P. Longwell, A.M. Dean, "Prediction of Rate Constants for Combustion and Pyrolysis by Bimolecular QRRK," AIChE J., 32, 1971 (1986).

[5] R.J. Kee, J.F. Grcar, M.D. Smooke, J.A. Miller, A Fortran Program for Modeling Steady Laminar One-Dimensional Premixed Flames, Report No. SAND85-8240, Sandia National Laboratories, 1985. Modification of PREMIX version 2.5, 1991.

[6] R.J. Kee, F.M. Rupley, J. A. Miller, Chemkin-II: A Fortran Chemical Kinetics Package for the Analysis of Gas-Phase Chemical Kinetics, Report No. SAND89-8009, Sandia National Laboratories, 1989.

See more of this Session: Poster Session: Computational Molecular Science and Engineering Forum
See more of this Group/Topical: Computational Molecular Science and Engineering Forum

Mechanistic Modeling of PLGA Microparticle Drug Delivery Systems

Ashlee N. Ford, Chemical & Biomolecular Engineering, University of Illinois, Urbana, IL

Daniel W. Pack, Chemical & Biomolecular Engineering, University of Illinois Urbana-Champaign, Urbana, IL

Richard D. Braatz, Department of Chemical and Biomolecular Engineering, University of Illinois at Urbana-Champaign, Urbana, IL

Controlled-release drug delivery systems are being developed as alternatives to conventional medical drug therapy regimens which require frequent administrations due to short pharmaceutical in vivo half-life and poor oral bioavailability. Controlled-release systems have the potential to provide better control of drug concentrations, reduce side effects, and improve compliance as compared to conventional regimens. The model-based design of controlled-release devices, such as biodegradable poly(lactic-co-glycolic acid) (PLGA) polymer microspheres, is challenging because of incomplete understanding of the mechanisms that regulate the release of drug molecules. This research focuses on modeling the autocatalytic polymer degradation and release of dispersed drug molecules from PLGA microspheres to capture size-dependent heterogeneous degradation behavior observed experimentally but not accounted for by existing models. Recently, other researchers have suggested that the autocatalytic polymer degradation is the primary mechanism by which the diffusive drug release is accelerated, and this process should depend strongly on particle size. The hypothesis of the present work is that simultaneously modeling the mathematics of diffusion, autocatalytic chemical reactions, chemical equilibria, and pore formation--the phenomena which are considered to contribute to the degradation of polymer particles and the release of drug molecules--rather than independently modeling any of the phenomena in a purely sequential manner will accurately mimic the actual overall release process. The developed mechanistic model tracks acid concentration as a function of space and time with a system of nonlinear partial differential equations for determination of intraparticle pH while modeling degradation kinetics, molecular weight distribution variation, and drug transport with varying diffusivity coupled to the concentrations of other reacting species. The chemical reaction mechanism including autocatalytic effects is coupled to a simplified diffusion model and pore formation model to incorporate spatial variations in degradation rate for all species within the microspheres. The model equations are solved for small computational time steps for an extended period of time in order to capture an entire release profile. The high resolution simulation of the coupling between reaction and diffusion captures important dynamical phenomena observed in experiments that cannot be modeled with the models in the literature that have simpler numerical solution but do not take the coupling into account. The inclusion of the spatial variation of autocatalytic effects is a unique contribution of this modeling work. The model for PLGA microspheres is extended to apply to core-shell microparticles and microcapsules, which are important options for encapsulating drugs for delivery in a multi-stage pulsatile release fashion or for protecting proteins from being deactivated by suspension in an aqueous core for time-delayed delivery after polymer microcapsule degradation. The predictions for the core-shell microparticles and microcapsules are compared to experimental results.

See more of this Session: Poster Session: Computational Molecular Science and Engineering Forum
See more of this Group/Topical: Computational Molecular Science and Engineering Forum

Ca^{2+}- Phospholipids Interactions: Insights Into Membrane Fusion Mechanisms

Zeena Kas Issa, Chemical Engineering, Wayne State University, Detroit, MI

Charles W. Manke, Chemical Engineering, Wayne State University, Detroit, MI

Bhanu P. Jena, Chemical Engineering, Wayne State University, Detroit, MI

Jeffrey J. Potoff, Chemical Engineering, Wayne State University, Detroit, MI

Prior x-ray diffraction, light scattering, photon correlation spectroscopy, and atomic force microscopy experiments suggest that SNARE-induced membrane fusion in cells proceeds as a result of calcium bridging opposing bilayers, which leads to the release of water from hydrated Ca^{2+} ions as well as the loosely coordinated water at PO-lipid head groups [1]. It is hypothesized that the combined effect of local dehydration of phosphate head groups, as a result of Ca^{2+} bridging, leads to a destabilization of lipid bilayers and resulting membrane fusion. This hypothesis was tested in the current study by performing atomistic molecular dynamic simulations in the isobaric-isothermal ensemble on dimethylphosphate anions (DMP^-) and Ca^{2+} in water at 298 K and 1.01 bar. Examination of molecular configurations obtained from the molecular dynamics simulations reveals extensive formation of DMP-Ca-DMP bridges. It was also observed that two Ca^{2+} are able to combine with two DMP^- to form a "ring complex" by bridging the anionic oxygens in DMP^-[2]. Average distances between Ca^{2+} and the anionic oxygens of bridged DMP^- are calculated as 2.92 Å, which is in close agreement with the 2.8 Å separation between vesicles reported by light scattering experiments of tv-SNARE induced membrane fusion[1]. Simulations also show that upon calcium bridging to DMP^-, water is removed from the anionic oxygens of DMP^- as well as the hydrated calcium ions.

The hypothesis that Ca^{2+} can bridge the head groups of apposed bilayers is tested with NPT molecular dynamics simulations of apposed bilayers containing either pure DMPC bilayers or mixed DMPC/POPS bilayers composed of 128 lipids of different compositions. Simulations are preformed with a 0.3 nm initial spacing between the bilayers, corresponding to the closet approach detected by light scattering experiments of tv-SNARE mediated vesicle fusion. Calcium bridging of apposing bilayers was observed after 10 ns. Additionally, simulations show structural changes in the lipid bilayers as well as the water molecules associated with the lipids upon calcium bridging the head groups.

[1] Jeremic, A., Cho, W-J., Jena, B.; J. Biol. Phys. & Chem. 4:139-142 (2004)

[2] Potoff, J., Issa, Z., Manke, C., Jena, B., Cell Biol. Int. 32:361-366 (2008)

See more of this Session: Poster Session: Computational Molecular Science and Engineering Forum
See more of this Group/Topical: Computational Molecular Science and Engineering Forum

Development of the TraPPE Force Field for Phospholipids

Navendu Bhatnagar, Chemical Engineering, Wayne State University, Detroit, MI

In this work, the TraPPE force field is extended to phospholipids, with the target application being the simulation of lipid bilayers and vesicles, and their interactions with various ions and solvents. A variety of methods for the calculation of partial charge distributions from quantum chemical calculations are considered. Parameters such as area per lipid, order parameters and bilayer thickness are calculated and compared to experimental values. In the case of DPPC, the area per lipid predicted by the TraPPE force field is 62.2 A2 which is close to the experimental value of 62.9 A2. Addition data are presented for DMPC, POPC and POPS.

See more of this Session: Poster Session: Computational Molecular Science and Engineering Forum
See more of this Group/Topical: Computational Molecular Science and Engineering Forum

Thermodynamics of Protein-Mediated Membrane Deformation - Application to Clathrin Dependent and Clathrin Independent Endocytosis

Neeraj J. Agrawal, Chemical and Biomolecular Engineering, University of Pennsylvania, Philadelphia, PA

Ravi Radhakrishnan, Department of Bioengineering, University of Pennsylvania, Philadelphia, PA

In eukaryotic cells, the internalization of extracellular cargo into the cytoplasm via the endocytosis machinery is an important regulatory process required for a large number of essential cellular functions, including nutrient uptake, cell-cell communication, and modulation of cell-membrane composition. Endocytosis is orchestrated by a variety of proteins. These proteins are implicated in membrane deformation/bending, cargo recognition and vesicle scission. While the involvement of these proteins have been established and their roles in membrane deformation, cargo recognition, and vesicle scission have been identified, current conceptual understanding falls short of a mechanistic description of the cooperativity and the bioenergetics of the underlying vesicle nucleation event which we address here using theoretical models based on an elastic continuum representation for the membrane and atomistic to coarse-grained representations for the proteins. We describe the energetics of deformations of membranes by using the Helfrich Hamiltonian represented in two different coordinate systems: the Monge Gauge and a Curvilinear coordinate system. The Monge approach is limited to small deformations of the membrane and thermal effects are included, however it cannot describe membrane shapes in the later stages of endocytosis during which membrane overhangs are observed. Thus we also employ the surface evolution approach to describe membrane geometries by minimizing the Helfrich Hamiltonian in a curvilinear coordinate system. This approach is versatile in describing membrane geometries in both small and large deformation limits but is limited to axis-symmetric profiles of membrane. Both approaches yield membrane energies. However to explicitly calculate the role of entropy change due to membrane bud formation, we employ thermodynamic integration method in conjugation with thermodynamic cycle which is possible in Monge formalism. In our model, curvature inducing proteins and protein assembly like epsin and clathrin coat affect the membrane Hamiltonian by changing the preferred mean curvature of the membrane. We model curvature induced by each epsin protein as a Gaussian function centered about the position of epsin protein. Clathrin assembly induces a constant curvature onto the membrane over a range dictated by the extent of clathrin coat polymerization. Diffusion of proteins over discrete membrane grids is handled through Monte-Carlo moves. We adopt a novel strategy of integrating two different phenomenological approaches, namely, time-dependent Ginzburg-Landau (TDGL) equation for membrane dynamics with Monte Carlo (MC) for protein diffusion. This integrated multiscale approach results in a unified description of membrane dynamics at the mesoscale (~μm, ~ms), under the influence of curvature-inducing proteins at the nanoscale (~10 nm, ~100 ns). Using this toolkit of the methods, we demonstrate the role of the endocytic protein assembly in driving membrane vesiculation and further quantify the energetics of the underlying process.

See more of this Session: Poster Session: Computational Molecular Science and Engineering Forum
See more of this Group/Topical: Computational Molecular Science and Engineering Forum

Virtual Screening of Quenchers for the Signal 3-Hydroxypalmitic Acid Methyl Ester in Ralstonia Solanacearum

Alejandro Alvarez, Chemical Engineering, Universidad de los Andes, Bogotá, Colombia

Adriana Bernal, Biology Science, Universidad de Los Andes, Bogotá, Colombia

Andrés F. González, Chemical Engineering, Universidad de Los Andes, Bogotá, Colombia

Ralstonia solanacearum, a widely distributed bacterial phytopathogen, causes a lethal wilting disease in more than 200 plants species. The expression of virulence and pathogenicity genes in R. solanacearum is controlled by a complex system that senses key environmental cues and triggers global shifts of gene expression. A central node in the control network is the quorum sensing (QS) system. Diverse kinds of QS signals have been reported in nature. Molecules in the group of N-acyl-homoserine lactones (AHLs) are the most commonly reported type of QS signals in bacteria. Although R. solanacearum exhibits an AHL-dependent system, the expression of virulence factors is AHL-independent. The chemical signal that controls the levels of virulence factors via quorum sensing in R. solanacearum is 3-Hydroxypalmitic acid methyl ester (3-OH PAME). In this work we used a computational approach to identify lipases with the potential ability to act as quenchers of the molecule 3-OH PAME. First we predicted de 3D structure of 3-OH PAME by geometry optimization using AMBER as a force field. Then we assess the capacity of the Candida antarctica's lipase B of being a quencher of 3-OH through molecular docking. To accomplish this, we evaluated the protein-ligand interaction between lipase B, N-Hexylphosphonate ethyl ester (Hee commun ligand) and 3-OH PAME. Molecular dynamics help to estimate the stability of the protein-ligand complexes obtained. Results obtained by this computational approach showed that lipase B could be a potential quencher of 3-OH PAME. We obtained through molecular dynamics a global minimum energy for 3-OH PAME of 3.827768 kcal/mol in contrast to its initial energy of 10.376802 kcal/mol. We also assess the kinetic and potential energies and the distance between atoms of 3-OH PAME and of the protein-ligand complex.

See more of this Session: Poster Session: Computational Molecular Science and Engineering Forum
See more of this Group/Topical: Computational Molecular Science and Engineering Forum

Estimation of Enthalpies of Formation of Biomolecules as Potential Biofuel Components: Comparison of Group Contribution and G3(MP2) Methods

Weisin Chai, Department of Chemical and Biomolecular Engineering, University of South Alabama, Mobile, AL

Kevin N. West, Department of Chemical and Biomolecular Engineering, University of South Alabama, Mobile, AL

James H. Davis Jr., Department of Chemistry, University of South Alabama, Mobile, AL

Interest in renewable energy sources has increased greatly in the past decade due to concerns about climate change and the cost of energy as well as national security interests in developing domestic energy sources. Biofuels offer an attractive alternative to fossil fuels as they offer a sustainable source of hydrocarbons, many of which are not competitive as food-stocks. Although several viable routes from biological materials to logistic fuels exist, there are a great number of potential biofeedstocks that are not presently used.

We have examined the heats of formation and heats of combustion of a variety of novel biologically derived molecules which may be viable biofuels, specifically as components of biodiesel. Traditionally, estimating heats of formation was only possible via one of a series of group contribution methods. However, for molecules dissimilar to those from the data set used to develop the group contribution theory parameters, these predictions were not always adequate. Modern computational capabilities and algorithms have allowed calculation of these gas-phase heats of formations from quantum chemical calculations combine with statistical mechanics. Such calculations do not suffer from the dependence on the training set as do group contribution methods, and as such can be applied more readily to newly synthesized or proposed compounds. In this study we compare estimates of the heats of formation for potential biofuels components from both group contribution methods and quantum chemical calculation methods (G3(MP2)) and compare both with available experimental results.

See more of this Session: Poster Session: Computational Molecular Science and Engineering Forum
See more of this Group/Topical: Computational Molecular Science and Engineering Forum

Anomalous and Single-File Diffusion for Fluids Confined in Microporous Carbons: Activated Carbon, Carbon Replicas, and Carbon Nanotube Systems

Joshua D. Moore, Department of Chemical & Biomolecular Engineering and Center for High Performance Simulation (CHiPS), North Carolina State University, Raleigh, NC

Keith E. Gubbins, Department of Chemical & Biomolecular Engineering and Center for High Performance Simulation (CHiPS), North Carolina State University, Raleigh, NC

Microporous carbons have found widespread use in industry due to their low production cost and high adsorption capacity. Disordered carbons, such as activated carbon, have been used extensively in environmental applications as adsorbents for the removal of toxic and odorous industrial chemicals. A second class of materials which has had recent interest are carbon templates of silica materials such as MCM-41, SBA-15 or various zeolites[1]. This type of material is inherently highly ordered. These materials have shown potential application in the areas of hydrogen storage and in the development of hydrogen fuel cells[2]. Realistic, atomically detailed models of these systems have previously been constrictued[3,4].

Using Grand Canonical Monte Carlo (GCMC) and equilibrium molecular dynamics (EMD) simulations, we have compared the adsorption and diffusion properties of fluids confined in realistic models of activated carbon with those of carbon replicas of zeolites. Both materials exhibit a maximum in the diffusion coefficient in the pore filling region of the isotherm as well as an anomalous (e.g., $t^{1/2}$ dependence of the mean squared displacement) for short times followed by a normal diffusive (Fickian) region. In order to explain the seemingly dual mode diffusion mechanism in these materials, we compare to diffusion of fluids confined in carbon nanotube systems (isolated carbon nanotubes and carbon nanotube bundles) which have distinct, bi-modal diffusion mechanisms.

See more of this Session: Poster Session: Computational Molecular Science and Engineering Forum
See more of this Group/Topical: Computational Molecular Science and Engineering Forum

Molecular Dynamics Studies On the Rheological Properties of Silica Nanoparticles Embedded Polyethylene Melt Using a Coarse-Grained Model

Yangyang Shen, Chemical and Biochemical Engineering, Rutgers University, Piscataway, NJ

The addition of nanoparticles to polymer composites has been shown to significantly influence the mechanical, optical, and electrical properties. We report a coarse-grained molecular dynamics study of silica nanoparticle aggregation and segregation in a polyethylene melt. Firstly a multiscale simulation scheme was developed for such system, where parameters of the solid-fluid interaction are determined from the results of a separate atomistic model. In our coarse grained model, every eight methylene groups of polyethylene are represented by one soft bead. The nanoparticles of 4nm in diameter are modeled as spherical clusters of beads kept together by rigid harmonic bonds. The particles move in simulation box of dimensions 30nm x 30nm x 30nm, and the simulations are carried out for 100ns.

Using this coarse-grained model, we explore the diffusion and rheological properties of polymeric fluids and nanoparticle composites under shear flow, and specifically we investigate the effects of polymer chain length, nanoparticle filling fraction and shear rate on viscosities η and first normal stress difference N_1. The shear viscosity is calculated from the tangential stress that is exerted on the walls. We demonstrate that the addition of nanoparticle fillers leads to a more pronounced shear thinning behavior, as a result of increasing shear rates in the gaps between the filler particles. The shear viscosity for longer molecules shows that, in Planar Couette Flow, shear thinning occurs independent of the molecular weight. Moreover, the addition of particles leads to a reduction in the first normal stress difference, which is commonly equivalent to the elasticity of the composite. These results are qualitatively consistent with the experimental studies of nonlinear rheology of particle-filled polymer melts.

In addition, we characterize the structure of the nano-composite using particle-particle radial distribution functions and the composite specific heat C_V. The dependence of the specific heat on the nanoparticle filling fraction exhibits a maximum occurs at about 3.6% wt, which we attribute to the nanoparticle agglomeration transition. At lower filling fractions the nanoparticles are in a dispersed state, and for filling fractions exceeding 3.6 wt%, the nanoparticles show a tendency to agglomerate. The polymer-mediated particle-particle forces exhibit a more repulsive character in the case of longer chains than in the case of the shorter ones. Moreover, the dispersed state of the nanocomposite becomes more stable as the molecular weight of polyethylene increases. This effect is most visible when the radius of gyration of the linear polymer exceeds the nanoparticle radius. The polymer gyration radius grows with nanoparticle filling fraction due to chain entanglement around the particles, which is typical in polymer reinforcement.

See more of this Session: Poster Session: Computational Molecular Science and Engineering Forum
See more of this Group/Topical: Computational Molecular Science and Engineering Forum

Mechanistic Dynamics of Single Chains in Dense Liquids Under Shear Flow

Jun Mo Kim, Chemical and Biomolecular Engineering, The University of Tennessee, Knoxville, TN

Brian Edwards, Chemical and Biomolecular Engineering, The University of Tennessee, Knoxville, TN

David Keffer, Chemical and Biomolecular Engineering, The University of Tennessee, Knoxville, TN

Bamin Khomami, Department of Chemical and Biomolecular Engineering, Material Research and Innovation Laboratory (MRAIL), University of Tennessee, Knoxville, TN

Many models have been developed to describe the bulk-averaged rheological and morphological properties of polymeric liquids. However they are still inadequate to describe many material properties at high strain rates. By tracking and comparing the motions of individual molecules as functions of strain rate, new ideas might emerge which will allow for the extension of existing models to relevant phenomena at atomic length and timescales. In this study, we focused on the mechanistic dynamics of chains over a wide range of shear rates. In the process of explaining the maximum value in the mean-squared, end-to-end vector vs. Weissenberg number (*Wei*) curve, we hypothesized that substantial rotational motion of individual chains at high *Wei* effectively lowers the magnitude of mean-squared, end-to-end vector: as the chains tumble through periodic cycles, they adopt hairpin-like configurations in which the chain ends pass very close to each other. This hypothesis also explains the disparate behavior of the probability distribution of the magnitude of end-to-end vector between low *Wei* and high *Wei*. At low *Wei*, the probability distribution displays a Gaussian behavior that is associated with the pre-averaged rheological theories. At high *Wei*, however, it is divided into two peaks, which are related to rotation and stretching of chains, respectively. The magnitude of end-to-end vector and orientation angle vs. time at high *Wei* depicts the details of the rotational or tumbling motion of individual chains and these two properties are correlated. We saw that the magnitude of end-to-end vector changed from a nearly fully-extended chain to tightly-coiled structure as functions of time, and that chain ends pass each other very closely during the tumbling cycle. From calculation of the time correlation of the components of end-to-end vector and its Fourier transformation, we extracted multiple timescales associated with dynamics of chain motion: these are associated with the frequency of rotation and the decay of the rotational correlation. These timescales will be very important factor to develop new model and to understand the response of complex polymeric liquids in the future.

See more of this Session: Poster Session: Computational Molecular Science and Engineering Forum
See more of this Group/Topical: Computational Molecular Science and Engineering Forum

First Principles Theoretical Analysis of Doping in II-VI Compound Semiconductor Nanocrystals

Tejinder Singh, Department of Chemical Engineering, University of Massachusetts, Amherst, MA

T. J. Mountziaris, Chemical Engineering, University of Massachusetts, Amherst, MA

Dimitrios Maroudas, Department of Chemical Engineering, University of Massachusetts, Amherst, MA

Nanocrystalline structures of compound semiconductors, such as the II-VI compounds ZnS, CdSe, and ZnSe, exhibit size-dependent optoelectronic properties and form the basis for a new generation of nanoelectronic and photovoltaic devices, as well as biological labels. Doping in semiconductor nanocrystals allows for precise control of their optical and electronic properties. However, in spite of its feasibility, doping of semiconductor nanocrystals has been an extremely difficult task. Self-purification of nanocrystals makes the incorporation of dopants into nanocrystals even more difficult. Typically, Mn-doped ZnSe nanocrystals have been grown using hot-injection organometallic synthesis. It has been observed that increase in anion:cation ratios leads to higher doping efficiencies. This has been attributed to adsorption of Mn dopant atoms into ZnSe nanocrystals through their surface facets, in particular ZnSe(001)-(2×1) surface facets. In this context, we aim at obtaining a fundamental and quantitative understanding of dopant adsorption and diffusion on II-VI semiconductor nanocrystal surfaces, which can help elucidate the mechanisms of dopant incorporation into growing nanocrystals.

In this presentation, we report results on dopant adsorption and diffusion on surface facets of ZnSe and ZnSe1-xSx nanocrystals based on first-principles density functional theory (DFT) calculations within the generalized gradient approximation. In our DFT calculations, we have employed slab supercells, plane-wave basis sets, and the projector-augmented wave method. We have also implemented the climbing-image nudged elastic band method to construct fully optimized dopant diffusion pathways. We have made indirect comparisons between our DFT analysis and experimental findings on doping efficiencies.

We have computed the surface energies of the three high-symmetry surfaces [(001), (110), and (111)] of ZnSe and examined the corresponding surface reconstructions and their stability as a function of the anion (Se) chemical potential. Based on these findings, we have constructed the equilibrium crystal shape (ECS) of ZnSe nanocrystals as a function of the Se chemical potential and found that anion-rich reconstructed surface facets are present in the ECS of a ZnSe nanocrystal. We have computed the binding energies of Mn dopant atoms onto all of the ZnSe nanocrystal facets and found that they depend strongly on the surface morphology and nanocrystal shape. We conclude that all anion-rich surfaces contribute toward dopant adsorption onto ZnSe nanocrystal surface facets.

In addition, our DFT calculations indicate that the binding energy for Mn adsorption onto various sites of all surface facets increases with increasing dopant surface concentration. This low binding energy at low dopant surface concentration explains the doping difficulties during nanocrystal growth. Furthermore, we have analyzed several dopant migration pathways for Mn diffusion on the ZnSe(001)-(2×1) surface. The calculated activation barriers for migration along the Se dimer rows range from 0.17 eV to 0.43 eV, depending upon the dopant surface concentration. Due to the low activation barriers, dopant atoms can migrate fast along the Se dimer rows without substantial surface relaxation. However, migration of the Mn atom across the Se dimer rows from the dimer site (Mn adsorbed onto the Se dimer) to the trough site (Mn adsorbed in the trough between Se dimers) is governed by a high-barrier pathway. Finally, the above analysis of dopant adsorption and diffusion on nanocrystal facets has been extended to ZnSe1-xSx nanocrystals in order to examine the effects of nanocrystal alloying on the doping efficiency.

See more of this Session: Poster Session: Computational Molecular Science and Engineering Forum
See more of this Group/Topical: Computational Molecular Science and Engineering Forum

Modeling of Bulk Metallic Glass Membranes for H2 Separation/Purification

Stephen L. Garrison, Computational Sciences, Savannah River National Laboratory, Aiken, SC

Kyle Brinkman, Materials Science & Technology, Savannah River National Laboratory, Aiken, SC

Paul Korinko, Materials Science & Technology, Savannah River National Laboratory, Aiken, SC

Thad Adams, Materials Science & Technology, Savannah River National Laboratory, Aiken, SC

Here we will present the results of our efforts using computational chemistry to model zirconium (Zr) based bulk metallic glasses (BMGs). The goal is to better understand the potential of BMGs as hydrogen gas separation or purification membranes and to guide experimental work for improving their membrane properties by appropriately modifying the alloying elements and relative compositions. Initial structures were developed using the efficiently-packed, solute-centered solvent cluster model with calculations performed using a planewave implementation of Density Functional Theory (DFT) in the Local Density Approximation (LDA). Investigations into issues such as diffusional paths, propensity towards hydride formation, and the dependence of these properties on the composition of the BMG will be discussed.

See more of this Session: First-Principles Simulation of Condensed Phases: Bulk Materials
See more of this Group/Topical: Computational Molecular Science and Engineering Forum

On the Mechanism of Proton Transport in Model Perfluorosulfonic Acid Systems: Ab Initio Molecular Dynamics Simulations

Bradley Habenicht, Chemical and Biomolecular Engineering, University of Tennessee, Knoxville, TN

Stephen Paddison, Chemical and Biomolecular Engineering, The University of Tennessee in Knoxville, Knoxville, TN

Perfluorsulfonic acid (PFSA) ionomers have drawn great interest in the scientific community due to their use as the electrolyte in proton exchange membrane fuel cells (PEMFCs). PEMFCs have the potential to provide an efficient and clean energy supply through the electrochemical conversion of hydrogen and oxygen into water. However, the mechanisms of proton transport through these membranes are poorly understood, particularly at low hydration levels. As these membranes are hydrated, a phase separation between the hydrophobic polymer backbone and the ionomeric, hydrophilic sidechains occurs, creating a labyrinthine, inhomogeneous system. Due to this complexity, as well as difficulties in ionomer synthesis, elucidating the structure-proton transport relationship has been difficult. A detailed understanding of this highly disordered system is required for the development of ionomers capable of facilitating efficient proton transport.

Experimental and theoretical efforts have shown that, upon hydration, PFSA ionomers swell, creating clusters and/or channels consisting of the aqueous domain and hydrated ions. At low hydration levels, these regions are quite small or narrow, being on the order of one nanometer in diameter. As the disorder of these channels makes them difficult to study theoretically, we have employed carbon nanotubes as scaffolding for model PFSA channels. Carbon nanotubes are very structurally defined on a nanometer length scale and their large variations in diameter provide a rich nanostructured architecture for the investigation of proton transport and diffusion, local environmental effects and protogenic spacing. Initial studies have indicated that side chain flexibility and motion of the sulfonate groups are important to efficient proton conductivity.

Ab Initio molecular dynamics (AIMD) is a computational technique which decouples the light, fast moving electrons from the slower heavier nuclei. The electrons are then treated quantum mechanically, while the nuclei are evolved classically, which saves much computational effort. As such, AIMD does not require empirical parameters, force fields, or assumptions about the mechanisms of proton transport. The former feature is extremely important, as AIMD may reveal surprising reaction dynamics and pathways. The AIMD simulations were performed using the Vienna Ab Initio Simulations Package (VASP). The simulation cells were constructed with periodic boundary conditions along the length of the CNT and 6 Å of vacuum in the perpendicular directions. The geometry was optimized to its minimum energy structure and the system was then heated to 300K using repeated velocity scaling. Once the cells were thermalized, 10 to 20 ps trajectories were obtained in the microcanonical ensemble for the calculation of diffusion coefficients and the analysis of ion formation.

See more of this Session: First-Principles Simulation of Condensed Phases: Bulk Materials
See more of this Group/Topical: Computational Molecular Science and Engineering Forum

Energetics of Proton Transfer of Hydrated Perfluorosulfonic Acids

Shenghong Zhang, Department of Chemical and Bimolecular Engineering, University of Tennessee, Knoxville, TN

Stephen Paddison, Chemical and Biomolecular Engineering, The University of Tennessee in Knoxville, Knoxville, TN

David Keffer, Chemical and Biomolecular Engineering, The University of Tennessee, Knoxville, TN

The hydrated proton exchange membrane (PEM) functions as the electrolyte in a fuel cell, effectively separating the anode and cathode and reactant gases. Hence, the conduction of protons is very important to the efficient operation of the device. Several experimental and computational studies have been undertaken to understand the mechanism of proton transport in these phase separated materials. A complete understanding of the mechanism has not been acquired but critical to long range transport is the formation of a continuous network of dynamic hydrogen bonds. In this work, the structure and energetics of proton transfer in minimally hydrated perfluorosulfonic acid (PFSA) membranes is explored through electronic structure calculations of model systems possessing the critical components.

PFSA membranes are the most commonly used electrolyte and electrode separator in PEM fuel cell due to their substantial chemical and mechanical stability. Under operating conditions, the acidic groups ($-SO_3H$) of the ionomer are hydrated and the protons dissociated generating protonic charge carriers. Proton may be transferred through structural diffusion via the solvating water molecules but also through coupled vehicular diffusion where there is a coupling of proton and water molecule as hydronium ions (H_3O^+). The hydrogen bonding of the water, water content, and density of the sulfonic acid groups is critical for the formation of the protonic charge carriers and mobility of the protons.

The energetics and associated structures for proton transfer in minimally hydrated PFSA membranes is investigated with electronic structure calculations of model systems consisting of pentafluoroethanesulfonic acid ($C_2F_5SO_3H$) hydrated with xH_2O ($x = 4, 5, 6, 7$) molecules or oligomeric fragments consisting of two or more perfluorinated side chains. Ab initio self-consistent field (SCF) molecular orbital calculations were performed using the GAUSSIAN 03 suite of programs. Several different initial configurations were chosen to avoid biasing the resulting fully optimized structure. The optimizations were undertaken without symmetry constraints initially employing Hartree-Fork theory with the 6-31G(d,p) basis set. The resulting equilibrium structures were then further refined with density functional theory with Becke's 3 parameter functional (B3LYP) and the same basis set. In the final optimized structure for each system, the hydrogen bond from the center oxygen (^1O) of each hydronium ion (H_3O^+) to the oxygen atom (^2O) in a neighboring H_2O was selected for transferring the proton between the oxygen atoms. The distance between ^1O and ^2O was constrained during the transfer and at least 2 different protons were transferred (separately) at each hydration level. These scanning calculations were performed at the B3LYP/6-31(d,p) level. The results show that the energy barrier of transferring a proton is strongly affected by the structure features of the neighboring waters, similar to what is observed in conversions between Zundel and Eigen ions in bulk water. However, the presence of the sulfonate group(s) tends to break the symmetry of the energy profile. The hydration level was also found to affect the energetic barrier in these systems.

See more of this Session: First-Principles Simulation of Condensed Phases: Bulk Materials
See more of this Group/Topical: Computational Molecular Science and Engineering Forum

Displacement Pathways in Bismuth Pyrochlores

Beverly Brooks Hinojosa, Department of Chemical Engineering, University of Florida, Gainesville, FL

Juan C. Nino, Department of Materials Science and Engineering, University of Florida, Gainesville, FL

Aravind R. Asthagiri, Department of Chemical Engineering, University of Florida, Gainesville, FL

There is considerable interest in pyrochlore systems ($A_2B_2O_7$) for use in high-permittivity dielectrics, capacitors, and high-frequency filter applications. The properties of these materials can be tuned through substitutions on the A and B cation sites, resulting in an extensive parameter space. Better understanding of the role of the local atomic structure and dynamics on the macroscopic properties will enable rational design within the vast number of possibilities. Using density functional theory (DFT), quantum mechanical calculations were performed to examine several Bi-containing pyrochlores with the Fd-3m (No. 227) space group to determine the role of chemical substitutions on the local geometric and electronic structure. We considered six simple bismuth pyrochlores ($Bi_2B^{4+}{}_2O_7$ with B^{4+} = Ti, Ru, Rh, Os, Ir, and Pt) and four pyrochlores with cation substitution selected based on availability of experimental IR and Raman spectroscopy ($Bi_{3/2}M^{2+}Nb_{3/2}O_7$ and $Bi_{3/2}M^{2+}Ta_{3/2}O_7$ with M^{2+} = Zn, Mg). We will present DFT results on the structural properties of the simple pyrochlores including equilibrium lattice constants, oxygen positional parameters, and atomic displacement patterns. For the pyrochlores with cation substitution, energetics related to cation ordering will be presented, as well as atomic displacement magnitudes and pathways. Where possible, our DFT results will be compared to available experimental studies to understand the role of the atomic substitutions on the material properties.

See more of this Session: First-Principles Simulation of Condensed Phases: Bulk Materials
See more of this Group/Topical: Computational Molecular Science and Engineering Forum

First-Principles Theoretical Analysis of Pure and Hydrogenated Crystalline Carbon Phases and Nanostructures

Tejinder Singh, Department of Chemical Engineering, University of Massachusetts, Amherst, MA

Michael J. Behr, Department of Chemical Engineering and Materials Science, University of Minnesota, Minneapolis, MN

Eray S. Aydil, Department of Chemical Engineering and Materials Science, University of Minnesota, Minneapolis, MN

Dimitrios Maroudas, Department of Chemical Engineering, University of Massachusetts, Amherst, MA

Carbon exists in a variety of different allotropes, such as cubic diamond, hexagonal diamond (lonsdaleite), graphite, carbynes, fullerenes, and carbon nanotubes. In addition to these well known forms, various other forms such as *i*-carbon, a cubic phase with lattice parameter $a = 4.25$ Å, a body-centered cubic (BCC) carbon phase with $a = 3.1$ Å, and *n*-diamond, a face-centered cubic (FCC) phase with $a = 3.57$ Å have been reported in experiments. These new carbon phases have been synthesized by various techniques, including plasma-assisted deposition from dilute hydrocarbon gases and hydrogen plasma exposure of bulk diamond and other carbon surfaces. However, the structure and stability of some of these new phases of carbon (*i*-carbon, *n*-diamond, and the new BCC phase) remain elusive.

In this presentation, we discuss a detailed analysis of various crystalline phases of carbon, which are candidate structures to explain the aforementioned experimentally observed carbon phases, as well as the potential role of hydrogen in determining the structures of the resulting carbon nanocrystals. Toward this end, we have employed first-principles density functional theory (DFT) calculations in order to compute the optimal lattice parameters and formation energies of pure, as well as hydrogenated crystalline phases of carbon. Our DFT calculations were carried out within the generalized gradient approximation and employed plane-wave basis sets, ultra-soft pseudopotentials, and supercell models.

We examined various crystalline cubic phases of pure carbon, including simple cubic, diamond-cubic, FCC, and BCC phases, as well as more complex crystalline phases of pure carbon such as supercubane, BC-8 carbon (the g-Si equivalent of carbon), and another BCC phase with space group I23 (#197), an optimal lattice parameter $a = 3.31$ Å, and a cohesive energy of 5.81 eV. We found that none of the pure carbon allotropes examined could explain satisfactorily the experimentally observed structures of *i*-carbon, *n*-diamond, or the new observed BCC carbon phase with $a = 3.1$ Å. Hence, we considered the incorporation of H in high-symmetry interstitial sites of diamond-cubic, BCC, and FCC carbon lattices, including bond-center, octahedral, and tetrahedral sites, and investigated the effect on the equilibrium lattice parameter of varying the composition, C_xH_y $(x,y \leq 3)$, of the carbon phases. H incorporation into the lattices of the various cubic carbon phases causes increased lattice expansion with increasing H concentration and enhances lattice stability compared to that of the corresponding pure carbon phases. As a result, by assuming interstitial H incorporation at varying concentrations in the cubic carbon lattices, we could provide possible interpretations for the structure of *i*-carbon, *n*-diamond, and the new BCC carbon phase.

See more of this Session: First-Principles Simulation of Condensed Phases: Bulk Materials
See more of this Group/Topical: Computational Molecular Science and Engineering Forum

Ab Initio Crystal Structure Prediction of Flexible Molecules

Andrei V. Kazantsev, Department of Chemical Engineering, Imperial College London, London, United Kingdom

Panos G. Karamertzanis, Department of Chemical Engineering, Imperial College London, London, United Kingdom

Claire S. Adjiman, Department of Chemical Engineering, Imperial College London, London, United Kingdom

Constantinos C. Pantelides, Process Systems Enterprise Ltd., London, United Kingdom

In previous work, we developed an algorithm for crystal structure prediction that identifies most low-energy minima (including the experimentally determined polymorphs) for crystals of both rigid [1] and flexible [2] molecules. This algorithm has been used successfully by our group and others to identify polymorphs for a wide range of molecules (including those with crystal structures in "unusual" space groups, and also more than one molecule in the asymmetric unit), as well as co-crystals and salts.

The main limitation of the above algorithm is the rapidly increasing computational cost as a function of molecular flexibility. This arises from the need to compute a priori a grid of intramolecular energies that can be used to construct restricted Hermite interpolants. Although the use of Hermite interpolants reduces the number of quantum mechanical calculations needed during the search, the dimension of the grid grows exponentially with the number of flexible degrees of freedom, practically limiting the applicability of the algorithm.

In this paper, we present substantial improvements of the algorithm that make the explicit consideration of a much larger number of flexible degrees of freedom tractable. The proposed approach is significantly more computationally efficient thanks to the use of local approximate models (LAMs) which are constructed based on a small number of quantum mechanical calculations performed "on-the-fly" during the search. The LAMs are used to compute the intramolecular energies based on a quadratic Taylor expansion of intra-molecular energy expressed in terms of the flexible molecular degrees of freedom. The expansion is constructed around a base point derived from an isolated-molecule wavefunction calculation of the intramolecular energy and its first and second order derivatives. The local approximate model is valid within a certain distance of the base point. The LAMs are generated as and when required and are stored in a look-up table, which allows their re-use if and when the search procedure returns to sample the same region of the conformational space.

The proposed algorithm requires only the atomic connectivity of the molecule under consideration and performs an extensive search for local minima of the lattice enthalpy surface by using deterministic low-discrepancy sequences to ensure an optimal coverage of the search space. Candidate structures can be generated in the 59 most common space groups with one or more crystallographically independent entities. These are then used as starting points of local optimization calculations performed with a sequential quadratic programming algorithm, which makes use of analytical partial derivatives with respect to all inter and intra- molecular degrees of freedom. A parallelized implementation of the algorithm allows minimizations from several hundreds of thousands of initial guesses to be carried out in reasonable time.

The validity of the proposed methodology is tested by predicting a set of single-component crystals and co-crystals containing flexible molecules. The use of the LAM significantly reduces the number of quantum mechanical evaluations that have to be performed. In practice, we find that many local optimizations can be done using a single quantum mechanical calculation carried out at the initial point. As a result, the proposed methodology addresses the computational limitations associated with our previous work and allows us to handle molecules with many flexible degrees of freedom.

References

[1] Karamertzanis, P. G. and Pantelides, C. C. J. Comput. Chem. 2005, 26, 304-324.

[2] Karamertzanis, P. G. and Pantelides, C. C. Mol. Phys. 2007, 105, 273-291.

See more of this Session: First-Principles Simulation of Condensed Phases: Bulk Materials
See more of this Group/Topical: Computational Molecular Science and Engineering Forum

3D Modeling of Wax Deposition in Oil/Water Stratified Flow

Zhenyu Huang, Chemical Engineering, University of Michigan, Ann Arbor, MI

H. Scott Fogler, The University of Michigan, Ann Arbor, MI

Wax deposition is a crucial economic issue in sub-sea oil production. Deposition modeling not only establishes scientific insight into deposition mechanisms, but also predicts deposit profiles for corrective methods such as pigging. Meanwhile, increasing off-shore drilling complicates wax deposition with multiphase flows. Previous research has been primarily focusing on single-phase wax deposition models in laminar/turbulent flow. The methods to calculate the hydrodynamics and heat/mass transfer under stratified flow are significantly different from well established single-phase flow correlations. Meanwhile, iterations on the momentum balance using computational fluid dynamics require extreme computational intensity. In this study, the first 3D deposition model in oil/water stratified flow has been developed based on the simplification of unidirectional flow in a laboratory scale flow loop. The unidirectional flow assumption helps to reduce dimensions of the calculations of the velocity profile, which is solved with successive over-relaxation (SOR) method. The 3D heat/mass transfer is solved with alternating-direction implicit (ADI) method. The model predicts that wax deposits on the upper part of the wall and reveals the effect of the addition of water. It is seen that the water phase, due to its greater heat capacity and greater thermal conductivity, has a significant impact on deposition by affecting the peak location of wax deposit, the cross-sectional averaged thickness and also in the azimuthal distribution of the deposits.

See more of this Session: Chemistry and Kinetics Integrated CFD Modeling
See more of this Group/Topical: Computational Molecular Science and Engineering Forum

FIRST PRINCIPLES BASED SIMULATIONS OF STEAM CRACKING OF ETHANE/PROPANE AND ETHANE/BUTANE MIXTURES

Maarten K. Sabbe, K.M. Van Geem, M.-F. Reyniers, G.B. Marin

Laboratory for Chemical Technology, Department of Chemical Engineering, Ghent University, Krijgslaan 281 S5, B-9000 Ghent, Belgium

Abstract

When simulating radical processes usually a combination of experimentally determined, predicted and fitted data is applied in order to overcome the lack of available kinetic and thermochemical data. This work presents the application of a consistent ab initio based kinetic and thermodynamic data in the simulation of radical reaction networks. The thermodynamic data are calculated using accurate quantum chemical CBS-QB3 calculations including corrections for hindered internal rotation, and the kinetics using CBS-QB3 based group additive models. Reactor simulations for steam cracking of ethane/propane/butane mixtures in pilot and industrial reactors over a wide range of process conditions show that, without adjusting any parameters, the main product yields can be predicted within 15% of the experimentally observed cracking yields.

Introduction

Design and optimization of industrial steam cracking reactors is focused on maximization of the product yields and the energy efficiency. An adequate kinetic description of the reaction chemistry in combination with an appropriate reactor model is essential for that purpose. This requires on the one hand that the chemistry involved in the steam cracking is modeled accurately at the molecular scale. On the other hand, the proposed reactor model should be of sufficient complexity to allow a proper description of the studied system. Only a fundamental simulation model with a reaction network based on elementary reaction steps, a so called single event microkinetic (SEMK) model, enables accurate and efficient modeling of the numerous reactions taking place in a steam cracking reactor. Even for cracking of light hydrocarbons, the networks describing the radical chemistry of the cracking process can contain up to thousands of reactions and hundreds of species. It is one of the most challenging tasks in reaction engineering to provide these networks with accurate kinetic and thermodynamic data. For steam cracking, the lack of thermodynamic and kinetic parameters for all elementary steps is commonly circumvented by using a combination of parameters originating from various sources such as kinetic databases (e.g. the NIST database), model predictions (e.g. Evans-Polanyi relation, group additivity schemes) and fitting to experimentally observed product yields. However, there is frequently a rather large scatter on the reported kinetic data for a given reaction. Also, the use of rate coefficients that are fitted to the observed product yields can obscure deficiencies in the reaction network. Therefore, in this work, all required thermodynamic and kinetic rate data are consistently obtained using group additive models based on high-accuracy CBS-QB3 quantum chemical calculations.

Methodology

Single event microkinetic model

The SEMK model for steam cracking consists of two parts: 1. the single event reaction network and 2. the reactor model equations and the solver of the resulting set of differential algebraic equations. The reaction network forms the heart of the SEMK model and determining the net reaction rates requires that both the thermochemistry and kinetics for each of the elementary steps in the reaction network are known. Calculating all required parameters by ab initio methods is infeasible due to the large number of reactions. Therefore, a group additive method is applied that links the thermodynamics and kinetics for larger species to high-accuracy ab initio data for smaller species.

Thermodynamic and kinetic data

The thermochemical values of primary importance for the calculation of equilibrium coefficients and enthalpy balances, are the standard enthalpy of formation $\Delta_f H°$(298 K), the standard entropy $S°$(298 K) and the molar heat capacity at constant pressure $C_p°(T)$, for hydrocarbons and hydrocarbon radicals. Thermodynamic data have been calculated using the CBS-QB3 method, involving corrections for all internal rotations. The applied methodology yields very good agreement with experimental data for standard enthalpies of formation (MAD=1.3 kJ mol^{-1}), entropies (MAD=1.2 J mol^{-1} K^{-1}) and heat capacities (MAD=1.2 J mol^{-1} K^{-1}). [1, 2] From these data a set of 95 consistent group additive values (GAVs) was derived for the calculation of the thermochemistry of large hydrocarbons and hydrocarbon radicals, involving 41 GAVs that had not been determined before.

To study the kinetics, the elementary reactions are grouped into reaction families. The 3 studied reaction families that contribute most significantly to the product yields in steam cracking are (i) radical addition to an unsaturated bond and the reverse β scission, (ii) hydrogen abstraction, and (iii) C-C and C-H bond scission and radical recombination reactions. Rate coefficients for the elementary steps have been calculated using the group additive method developed by Saeys et al.[3] extended to pre-exponential factors.[4, 5] In the group additive method, the activation energy and the pre-exponential factor are related to the structure of the transition state, which is expressed by adding group additive values ΔGAV° to the Arrhenius parameters of the reference reaction. The rate coefficient k is written as function of the tunneling coefficient κ, the number of single events n_e and the single-event rate coefficient $\tilde{k}$:

$$k = \kappa n_e \tilde{k} = \kappa n_e \tilde{A} \exp(-\frac{E_a}{RT}) \tag{1}$$

$$E_a(T) = E_{a,\text{ref}}(T) + \sum_{i=1}^{n} \Delta GAV^{\text{o}}_{E_a}(\text{C}_i) \tag{2}$$

$$\log A(T) = \log \tilde{A}_{\text{ref}}(T) + \sum_{i=1}^{n} \Delta GAV^{\text{o}}_{\log \tilde{A}}(\text{C}_i) + \log n_e \tag{3}$$

in which $E_{a,\text{ref}}$ and $\tilde{A}_{\text{ref}}$ are the Arrhenius parameters of the reference reaction, the $\Delta GAV°$ are the group additive values and the number of single events n_e is defined as:

$$n_e = \frac{n_{\text{opt},\ddagger}}{\prod_j n_{\text{opt},j}} \frac{\prod_j \sigma_j}{\sigma_\ddagger} \tag{4}$$

with n_{opt} the number of optical isomers, σ the global symmetry number and j running over the number of reactants. The use of group additive values relative to a reference reaction to calculate the activation energy and the pre-exponential factor of radical reactions results from the strong analogy that exists between the different reactions, considering their mechanism.[6]

For any reversible elementary chemical reaction, it follows from thermodynamic consistency that:

$$K = \frac{k_{forward}}{k_{backward}} \qquad \text{with } K = \exp\left(\frac{-\Delta_r G^0}{RT}\right) \quad \text{and} \quad k = A\exp\left(\frac{-E_a}{RT}\right) \tag{5}$$

Based on this equation, the reverse rate coefficients are calculated based on the forward rate coefficients and the thermodynamic equilibrium coefficient, which is calculated using the ab initio thermochemistry discussed above.

The group additive values are determined from kinetic parameters calculated using the CBS-QB3 method taking into account the internal rotation about the forming/breaking bond. Rate coefficients are obtained using the conventional transition state theory with corrections for tunneling when necessary:

$$k_\infty(T) = \kappa(T)\frac{k_B T}{h}\frac{n_{\text{opt},\ddagger}q_\ddagger}{n_{\text{opt},A}q_A n_{\text{opt},B}q_B}e^{-\frac{\Delta E(0\text{ K})}{RT}} \tag{6}$$

Out of several ab initio methods, this method gave the best agreement in comparison with experimental rate coefficients for a test set of 48 reactions containing all the reaction families relevant to steam cracking. The CBS-QB3 method with hindered rotation and Eckhart tunneling corrections yielded the lowest mean factor of deviation from the experimental rate coefficients, smaller than a factor 3 in most cases which is better than many commonly used DFT-based methods, see Table 1.[7, 8]

Table 1: Mean factors of deviation <ρ> between calculated and experimental rate coefficients (CBS-QB3 calculated rate coefficients with correction for internal rotation about the forming/breaking bond, and inclusion of tunneling correction for hydrogen addition and abstraction).

T (K)	Carbon radical[7]		Hydrogen radical[5]		Hydrogen[8]	Mean deviation
	addition	β scission	addition	β scission	abstraction	<ρ>
300	2.1	3.3	1.7	2.3	10.0	3.9
600	3.2	2.5	1.5	2.2	3.9	2.7
1000	3.9	2.6	2.0	1.8	5.8	3.2
Nr. of reactions in test set	8	6	7	6	21	48

Reactor simulation

Steam cracking is a non-isothermal, non-adiabatic and non-isobaric process. Hence, the 1 dimensional reactor model equations consist of the transport equations for mass, momentum and energy. The steady state continuity equation for a SEMK component j in the process gas mixture over an infinitesimal volume element with cross sectional surface area Ω, circumference ω and length dz is:

$$\frac{dF_j}{dz} = \left(\sum_{k=1}^{n_r} \upsilon_{kj}\, r_k\right)\Omega \tag{7}$$

with F_j the molar flow rate of SEMK component j, r_k the reaction rate of reaction k, and υ_{kj} the stoechiometric coefficient of SEMK component j. The energy equation is given by:

$$\sum_j F_j c_{pj} \frac{dT}{dz} = \omega\, q + \Omega \sum_k r_k (-\Delta H_k) \tag{8}$$

with q the heat flux to the process gas, c_{pj} the heat capacity of component j at temperature T, $\Delta_f H_k$ the standard enthalpy of formation of species k, $R_{v,k}$ the net production rate for species k. The momentum equation accounting for friction and changes in momentum is given by:

$$-\frac{dp_t}{dz} = \alpha \left(\frac{2f}{d_t} + \frac{\zeta}{\pi\, r_b} \right) \rho_g\, u^2 + \alpha\, \rho_g\, u \frac{du}{dz} \tag{9}$$

with p_t the total pressure, α a conversion factor, f the Fanning friction factor, ρ the density of the gas mixture, r_b the radius of the bend, d_t the diameter and v the velocity. The boundary conditions of the model equations are:

$$T=T_0 \qquad C_j=C_{j0} \qquad p=p_0 \qquad (z=0) \tag{10}$$

Integration of the last two model equations is only required when the temperature and/or pressure profile are not imposed. Based on the reactions, and reaction rate coefficients, the production rate $r_{v,k}$ of each component j by reaction k, can be expressed as a function of the concentrations of the involved species. The resulting set of continuity equations forms a stiff set of non-linear first order differential equations. The stiffness of the set of equations originates from the large difference (several orders of magnitude) of the eigenvalues related to the molecular species on the one hand and the radical species on the other hand. The stiff solver DASSL (Li and Petzold, 1999) is used to integrate the stiff set of differential equations.

Based on the thermodynamic and kinetic data calculated in this work, a reaction network was automatically generated [9, 10]. The reaction mechanism generated using only the ab initio thermodynamic and kinetic data has already shown to yield excellent results for ethane cracking[11] and enabled, for the first time, to simulate an industrial steam cracker on an entirely ab initio basis without using a single adjustable parameter or experimental value. The proposed methodology is validated against pilot and industrial data for the co-cracking of ethane, propane and butane

Results

Pilot plant experiments

The experimental pilot plant set-up of the Laboratory for Chemical Technology allows fundamental studies of the kinetics of the cracking reactions[12] as well as of practical issues such as coke deposition in both the radiant coil[13] and the transfer line exchanger (TLE).[14] The pilot plant installation consists of 3 parts: a feed section, the furnace with the suspended reactor coil and the analysis section. The tubular reactor used in this set of experiments is made of Incoloy 800H, has a length of 23.14 m and an internal diameter of 10^{-2} m. These dimensions are chosen to achieve fast turbulent mixing in the coil at the applied feed flow rates. The temperature profile along the reactor is measured and regulated by fired burners.

The generated reaction network is used to simulate 12 pilot plant experiments, with the coil outlet temperature (COT) varying between 1004 and 1121 K and de coil outlet pressure, COP, between 0.152 and 0.296 MPa. The total hydrocarbon flow rate is between 3.1 and 4.0 kg/h and the feed consists of mixed compositions of ethane (10-68%), propane (19-77%) and *n*-butane (0-24%), with traces of methane (<3% for all experiments).

The pilot plant experiments were performed in a broad range of experimental conditions. Using the first principles based reaction network, the conversions are well reproduced, as can be observed from the yields of the reactants in the left panel in Figure 1. The main product yields (right panel of Figure 1) could be reproduced within 15%. There are also significant simulated 1,3-butadiene and benzene yields, up to 1.7 and 1.3 w% resp., but these yields are not shown ince these yields were not measured for most experiments. The remaining product yields are all smaller than 1 w%.

Table 2: Comparison between ab initio simulated product yields (wt%) and pilot plant data. Conditions: F_0 = 3.57 kg h^{-1}, 10% ethane, 77.4% propane, 12.6% *n*-butane, no steam dilution, COT = 1123 K, COP = 0.16 MPa.

component	Product yield (wt%)	
	Experiment	Ab initio prediction
H_2	1.40	1.46
methane	18.95	17.88
ethyne	0.34	0.83
ethene	32.90	31.73
ethane	8.11	8.16
propyne	0.37	0.29
1,2-propadiene	0.00	0.43
propene	16.70	18.14
propane	14.75	15.23
1,3-butadiene	2.17	1.66
1-butene	0.88	0.20
2-butene	0.38	0.13
isobutene	0.14	0.07
isobutane	0.04	0.09
n-butane	0.78	1.00
cyclopentadiene	0.57	0.39
benzene	1.12	1.31

The simulation results for the cracking of a single ethane/propane/butane mixture are given in Table 2. As discussed previously, the product yields are predicted within 15%, except for the minor product yields. The overprediction of the ethyne and 1,2-propadiene yields is caused by a slightly too fast β scission of vinylic and propargylic radicals, which has its origin in fast hydrogen addition reactions due to the neglect of pressure dependence.

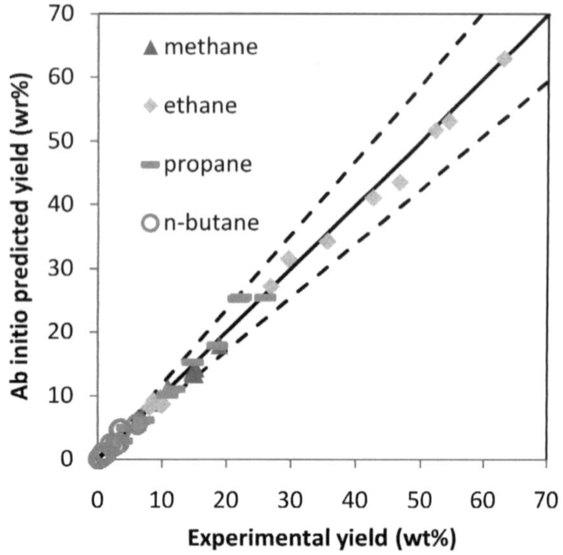

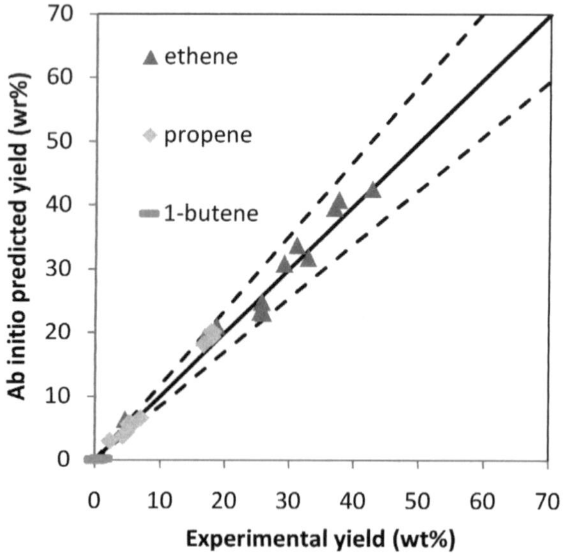

Figure 1: Ab initio predicted versus experimental product yields for the feedstock components (left) and the main products (right) ethene, propene and 1-butene in the pilot plant simulations. The conditions are given in the text. [--- 15% rel. interval]

Industrial reactor

The simulation is performed for the radiation section of an industrial steam cracker. The details of the reactor equations and integrating routines can be found in Van Geem et al.[15] The hydrocarbon feed rate is 0.534 kg/s (60% ethane and 40% propane) with a steam dilution of 0.3 kg steam/kg feed. The COT is 1125 K and the COP 0.21 MPa. The results in Figure 2 show that both conversion and selectivities are accurately predicted. Only for 1,3-butadiene, a component that is difficult to measure experimentally, a slightly larger deviation is observed.

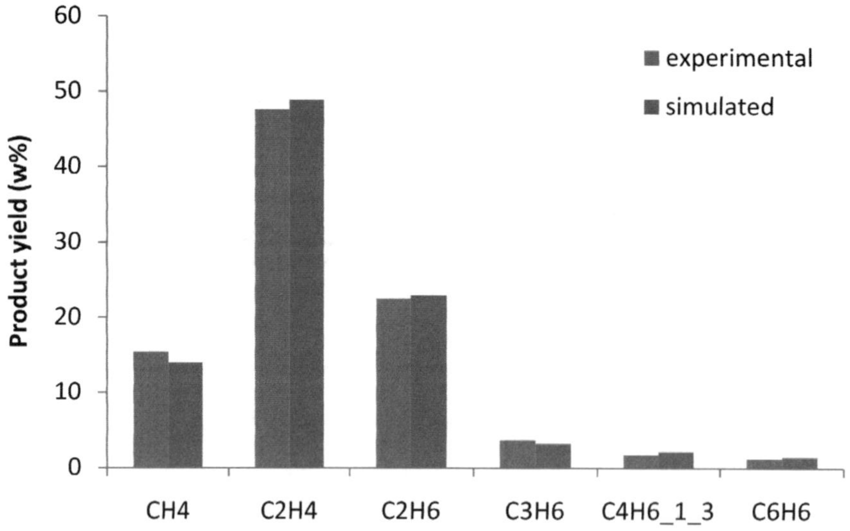

Figure 2: Experimental vs. ab initio predicted product yields for an industrial steam cracker ($F°$=0.534 kg hydrocarbon/s of which 60% ethane and 40% propane, δ= 0.3 kg steam/kg feed, COT=1125 K, COP=0.21 MPa).

Conclusions

This work presents the application of a first-principles based group additive method for the calculation of thermodynamics and Arrhenius parameters involved in the radical chemistry of hydrocarbons. The consistently developed group additive model is applied in the simulation of the steam cracker of ethane/propane/butane mixtures, in contrast to the usually applied combination of experimentally determined, predicted and fitted data. Using only these ab initio based additive models, the main product yields for a pilot and industrial cracker can be simulated within 15%. In the knowledge that any parameter has been adjusted and no experimental values have been used, the agreement can be considered remarkable. The remaining deviations might be caused by pressure dependence of the rate coefficients, which is currently neglected. Nevertheless, these results illustrate the potential of using ab-initio calculated thermodynamic and kinetic data for the simulation of a chemical processes.

Acknowledgements

M.K.S. held a Ph.D. grant of the Institute for the Promotion of Innovation through Science and Technology in Flanders (IWT-Vlaanderen) and is grateful for financial support from the Fund for Scientific Research-Flanders (F.W.O.-Vlaanderen). K.M. Van Geem holds a Postdoctoral Fellowship of the Fund for Scientific Research, Flanders, Belgium and a Fulbright fellowship for performing postdoctoral research at the Massachusetts Institute of Technology.

References

[1] M. K. Sabbe, M. Saeys, M. F. Reyniers, G. B. Marin, V. Van Speybroeck, M. Waroquier. *J. Phys. Chem. A* **2005**, *109*(33), 7466-7480.

[2] M. K. Sabbe, F. De Vleeschouwer, M. F. Reyniers, M. Waroquier, G. B. Marin. *J. Phys. Chem. A* **2008**, *112*(47), 12235-12251.

[3] M. Saeys, M. F. Reyniers, G. B. Marin, V. Van Speybroeck, M. Waroquier. *AIChE J.* **2004**, *50*(2), 426-444.

[4] M. K. Sabbe, M. F. Reyniers, V. Van Speybroeck, M. Waroquier, G. B. Marin. *ChemPhysChem* **2008**, *9*(1), 124-140.

[5] M.K. Sabbe, M.F. Reyniers, M. Waroquier, G.B. Marin. Hydrogen radical addition to unsaturated hydrocarbons and reverse beta-scission reactions: modeling of activation energies and pre-exponential factors. ChemPhysChem, in press, 2009.

[6] F. O. Rice, K. F. Herzefeld. *J. Am. Chem. Soc.* **1934**, *56*, 284-289.

[7] M. K. Sabbe, A. G. Vandeputte, M. F. Reyniers, V. Van Speybroeck, M. Waroquier, G. B. Marin. *J. Phys. Chem. A* **2007**, *111*(34), 8416-8428.

[8] A. G. Vandeputte, M. K. Sabbe, M. F. Reyniers, V. Van Speybroeck, M. Waroquier, G. B. Marin. *J. Phys. Chem. A* **2007**, *111*(46), 11771-11786.

[9] W. H. Green. *Adv. Chem. Eng.* **2007**, *32*, 1-50.

[10] K. M. Van Geem, M. F. Reyniers, G. B. Marin, J. Song, D. M. Matheu, W. H. Green. *AIChE J.* **2006**, *52*(2), 718-730.

[11] Van Geem, K. M., Sabbe, M. K., Reyniers, M. F., and Marin, G. B. First principles based simulations of steam cracking of light hydrocarbons. 20th International Symposium on Chemical Reaction Engineering. 7-9-2008. Kyoto.

[12] S. Wauters, G. B. Marin. *Ind. Eng. Chem. Res.* **2002**, *41*(10), 2379-2391.

[13] M. F. S. G. Reyniers, G. F. Froment. *Industrial & Engineering Chemistry Research* **1995**, *34*(3), 773-785.
[14] I. Dhuyvetter, M. F. Reyniers, G. F. Froment, G. B. Marin, D. Viennet. *Ind. Eng. Chem. Res.* **2001**, *40*(20), 4353-4362.
[15] K. M. Van Geem, G. J. Heynderickx, G. B. Marin. *AIChE J.* **2004**, *50*(1), 173-183.

Building Kinetic Models for the Liquid Phase: Hydrocarbon Autoxidation

Richard H. West, Amrit Jalan, and William H. Green
Massachusetts Institute of Technology, 77 Massachusetts Avenue 66-270, Cambridge MA 02139

Abstract

We present new methods for automatically generating and solving detailed kinetic models for liquid phase reactive systems, including estimates of solvation thermodynamics and diffusion kinetics, and apply this to the autoxidation of a diesel fuel surrogate.

Introduction

Kinetic models for gas-phase reacting systems, such as atmospheric chemistry and combustion, often contain thousands of species and reactions. Researchers in these fields have developed a number of tools to help them generate [1–5], manage [6], and solve [7–9] these detailed models. Many reacting systems of interest to chemical engineers, however, occur in the liquid phase. Liquid phase kinetic models of similar complexity, and the tools required to work with them, have yet to be developed. To begin, we have modified our Reaction Mechanism Generator (RMG) software [4, 5] enabling it to estimate the thermodynamics of solvation for reacting species using Abraham's model [10] (a Linear Solvation Energy Relationship) with parameters from a group contribution method formulated by Platts *et al.*[11]. This allows RMG to automatically generate kinetic models with solvation-corrected thermochemistry for reactions in solution. Secondly, we have developed a kinetic model solver that incorporates diffusion limited reaction rates, estimating species diffusivities using the Stokes-Einstein Relation, so that detailed kinetic models can be easily used in simulations of liquid phase systems. Finally, we demonstrate the use of these tools to investigate the liquid phase autoxidation of a mixture of hydrocarbons representing diesel fuel.

Automatic kinetic model generation with RMG

RMG is an automatic reaction mechanism generator [4, 5, 12, 13]. Given some starting species (e.g. methane and oxygen) and some reaction conditions (temperature and pressure) it will create a kinetic model of the reaction mechanism consisting of many (up to thousands) elementary reactions between intermediate species. Inside the software molecules are represented as graphs, with atoms as nodes and bonds as edges connecting the nodes. Standard graph-theory methods are used to identify equivalent graphs and ensure uniqueness. RMG uses "reaction families" to generate all the possible reactions that a species can undergo in the presence of the other species in the chemical mechanism. Every reaction family represents a particular type of elementary chemical reaction, such as bond-breaking, or radical addition to a double bond. Each reaction family has a recipe for mutating the graph, and a library of rate expressions for different reacting sites.

Because the model can contain thousands of species and rates, the estimation of thermochemical and kinetic parameters must be very fast. As with most mechanism generating tools, RMG uses a database of known values wherever possible to find thermochemical data for species, but usually the data are unknown and it estimates parameters using a group contribution method. Thermochemistry estimates are based on Benson's group additivity method for standard enthalpies of formation [14, 15]. In this additivity method a group is defined as a polyvalent atom together with all of its ligands. Each group (e.g. "a carbon that is bonded to one CO group, one carbon that has a double bond, and two hydrogens") contributes a set amount to the overall enthalpy of formation. The functional groups are recognized using a graph-theory matching algorithm. A similar method is used to estimate the rate coefficients for the reactions.

RMG uses a rate-based termination criterion; the reaction network is expanded until the rates of all reactions going to species not included in the network fall below a certain threshold. This helps to include important pathways without unnecessarily exploring slower pathways, rather than terminating the expansion after a set number of generations [16].

Automatic solution phase corrections

For liquid phase simulations the estimated thermochemistry of a species must be modified to take into account its solvation. Using equilibrium thermodynamics, the free energy of solvation ($\Delta G^{\circ}_{\text{solv}}$) is related to the concentrations of a species in the solvent phase and gas phase at equilibrium, by

$$\Delta G^{\circ}_{\text{solv}} = -RT \ln \left(c_{\text{solvent}}/c_{\text{gas}}\right) = -2.303RT \log_{10} K \tag{1}$$

where $K = \left(c_{\text{solvent}}/c_{\text{gas}}\right)$ is known as the partition coefficient.

We use Abraham's model, which, given a set of five molecular descriptors for a solute, can estimate the partition coefficient of the solute between different phases for which solvent descriptors are also known [10]. In the Abraham model, the partition coefficient of a solute between the gas phase and a solvent phase is given by

$$\log_{10} K = c + eE + sS + aA + bB + lL \tag{2}$$

where the upper case parameters (A, B, E, L, S) are solute descriptors (unique to a given solute) and the lowercase parameters (c, e, s, a, b, l) are solvent descriptors. The descriptor definitions have been chosen to reduce correlation between them and to try to make them physically meaningful. For example, the eE and sS terms mostly account for electrostatic interactions like dipole–dipole and dipole–induced-dipole, the aA and bB terms account for hydrogen bonds from the solute to the solvent and *vice versa*, and the lL term represents the contributions of cavity formation and dispersion.

Because of these efforts to make the solute descriptors physically meaningful, they can be predicted with reasonable success from the molecular structure of the solute. This can be done using the group contribution method of Platts *et al.*[11]. The idea is similar to Benson's group additivity method for standard enthalpies of formation: each group in the species contributes a set amount to each of the solute descriptors for that species.

Using Platts's group contribution scheme to provide solute parameters to the Abraham model, the Abraham model (2) to predict the partition coefficient between solvent and vapor, and the equilibrium equation (1) to link partitioning with free energy difference, we are able to predict the Gibbs free energy of solvation for any species at 298 K.

The Abraham model predicts partition coefficients at 298 K. To model systems at different temperatures, we need to estimate $\Delta G^{\circ}_{\text{solv}}$ as a function of temperature. As a first approximation, observing that

$$\Delta G^{\circ}_{\text{solv}} = \Delta H^{\circ}_{\text{solv}} - T\Delta S^{\circ}_{\text{solv}}, \tag{3}$$

we split $\Delta G^{\circ}_{\text{solv}}$ into contributions from solvation enthalpy $\Delta H^{\circ}_{\text{solv}}$ and solvation entropy $\Delta S^{\circ}_{\text{solv}}$, and assume that these do not vary with temperature.

To estimate the split, we assume that the entropic part of $\Delta G^{\circ}_{\text{solv}}$ is all due to cavitation (making a cavity in the solvent and placing the solute molecule in it), and that cavitation does not contribute to $\Delta H^{\circ}_{\text{solv}}$ [17]:

$$\Delta S^{\circ}_{\text{solv}} = \Delta S^{\circ}_{\text{cavitation}} = -\left(\frac{\partial \Delta G^{\circ}_{\text{cavitation}}}{\partial T}\right)_P \tag{4}$$

We estimate $\Delta G^{\circ}_{\text{cavitation}}(T)$ using the scaled particle theory method described by Pierotti [18]. This method requires the density of the solvent and the radii of both solvent and solute. The solvent properties are entered once per simulation; the radius of each species (as a solute) is estimated by assuming it to be spherical and using the UNIFAC group contribution scheme for Van der Waals volume [19, 20].

Together, this combination of approximations and estimates (fig. 1) allows us to estimate quickly the thermochemistry of an arbitrary species in solution, provided that the species is made up of atoms for which group values have been derived and that the solvent properties are known. We have automated this scheme and implemented it inside the reaction mechanism generation software RMG, so that these modified solution-phase thermochemical data are used when evaluating the reaction rate and equilibrium constants.

Solving detailed kinetic models in the solution phase

Simulating a reacting chemical system with a detailed kinetic model requires solving a list of ordinary differential equations (ODEs) – one for each species concentration. The terms in these ODEs represent the chemical reactions that create or consume the species, and their magnitudes depend on species concentrations and reaction rate coefficients. For reverse reactions these rate coefficients are calculated using the equilibrium constant, which is in turn dependent on the thermochemistry of the reacting species. Although any stiff ODE solver can be used to integrate the ODEs, the complexity is in managing and correctly evaluating all of the terms in the equations. Software to take care of these details and simplify running kinetic simulations in the gas phase has been around for decades (e.g. Chemkin [7] and the open-source Cantera [9]) but reactions in solution require additional considerations.

Reaction rate coefficients in gas phase kinetic simulations are usually represented by modified Arrhenius expressions ($k_r = AT^n \exp(-E_a/RT)$). In solution, however, the rate is also a function of

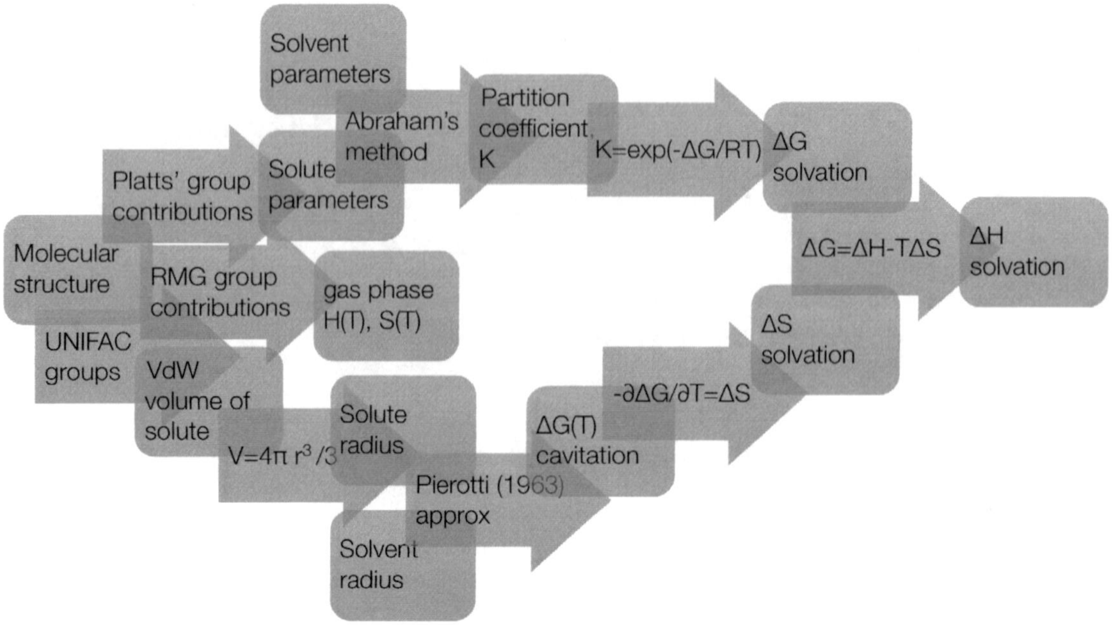

Figure 1: *Overview of method for estimating solvation thermochemistry from molecular structure.*

(and is often limited by) the rate of diffusion of the reacting species towards each other (or products away from each other). A simple model for this is based on Smoluchowski theory [21]:

$$k_{\text{eff}} = \frac{4\pi R\mathcal{D}k_r}{4\pi R\mathcal{D} + k_r} \tag{5}$$

where R is the encounter radius, $\mathcal{D}$ is the diffusivity of the reacting species towards each other, and k_r is the intrinsic rate coefficient without diffusion effects.

In RMG we estimate species diffusivities using the Stokes-Einstein relation. For this we need the diameter of each species, which we again estimate using the UNIFAC group contribution scheme [19, 20], and the viscosity of the solvent.

Modeling surrogate diesel fuel autoxidation

The tools described above were used to investigate the autoxidation of a diesel fuel surrogate when it is exposed to air and heated. Both the changes in thermochemistry due to solvation, and the changes in reaction rates due to diffusion, altered the overall predicted rate of autoxidation.

We hope that these new, free, tools for generating and using detailed kinetic models of reacting systems in solution will be of general interest to a wide audience and find a variety of uses in the chemical engineering community.

References

[1] Broadbelt LJ, Stark SM, Klein MT. Computer Generated Pyrolysis Modeling: On-the-Fly Generation of Species, Reactions, and Rates. *Ind Eng Chem Res*. 1994;33(4):790–799. doi:10.1021/ie00028a003.

[2] Grenda J, Androulakis I, Dean A, Green W. Application of computational kinetic mechanism generation to model the autocatalytic pyrolysis of methane. *Industrial and Engineering Chemistry Research*. 2003;42(5):1000–1010. doi:10.1021/ie020581w.

[3] Broadbelt L, Pfaendtner J. Lexicography of Kinetic Modeling of Complex Reaction Networks. *AIChE J*. 2005;51(8):2112–2121. doi:10.1002/aic.10599.

[4] Song J, Raman S, Yu J, Wijaya C, Stephanopoulos G, Green W. RMG: the next generation of automatic chemical reaction mechanism generator. In: *Proceedings of the AIChE Annual Meeting*. San Francisco, CA, USA. 2003; .

[5] RMG open source project home page. 2009.
URL http://rmg.sourceforge.net

[6] Myers J, Allison T, Bittner S, Didier B, Frenklach M, Green W, Ho Y, Hewson J, Koegler W, Lansing C, et al. A collaborative informatics infrastructure for multi-scale science. *Cluster Computing*. 2005;8(4):243–253. doi:10.1007/s10586-005-4092-4.

[7] Kee RJ, Rupley FM, Miller JA. Chemkin-II: A Fortran Chemical Kinetics Package for the Analysis of Gas-Phase Chemical Kinetics. *Tech. Rep. SAND89-8009*, Sandia National Laboratories. 1989.

[8] Lutz AE, Kee RJ, Miller JA. Senkin: a Fortran program for predicting homogeneous gas phase chemical kinetics with sensitivity analysis. *Tech. Rep. SAND87-8248 Revised*, Sandia National Laboratories, Livermore, Sandia National Laboratories, Livermore, CA 94551-0969. 1997.
URL http://public.ca.sandia.gov/chemkin/docs/senkin.pdf

[9] Goodwin DG. An open source, extensible software suite for CVD process simulation. In: *Chemical Vapor Deposition XVI and EUROCVD 14*, edited by Allendorff M, Maury F, Teyssandier F, vol. 2003-08 of *ECS Proceedings*. The Electrochemical Society, Pennington, New Jersey, USA: The Electrochemical Society. 2003; pp. 155–162.
URL http://www.cantera.org

[10] Abraham MH. Scales of solute hydrogen-bonding: their construction and application to physicochemical and biochemical processes. *Chem Soc Rev*. 1993;22(2):73–83. doi:10.1039/CS9932200073.

[11] Platts JA, Butina D, Abraham MH, Hersey A. Estimation of Molecular Linear Free Energy Relation Descriptors Using a Group Contribution Approach. *J Chem Inf Comput Sci*. 1999; 39(5):835–845. doi:10.1021/ci980339t.

[12] Song J. Building Robust Chemical Reaction Mechanisms: Next Generation of Automatic Model Construction Software. Ph.D. thesis, Massachusetts Institute of Technology, Cambridge

MA, USA. 2004.
URL http://hdl.handle.net/1721.1/30058

[13] Van Geem K, Reyniers M, Marin G, Song J, Green W, Matheu D. Automatic reaction network generation using RMG for steam cracking of n-hexane. *AIChE J*. 2006;52(2):718–730. doi:10.1002/aic.10655.

[14] Benson SW. *Thermochemical Kinetics*. New York: Wiley, 2nd ed. 1976.
URL http://www.worldcat.org/oclc/2074434

[15] Cohen N, Benson S. Estimation of heats of formation of organic compounds by additivity methods. *Chem Rev*. 1993;93(7):2419–2438. doi:10.1021/cr00023a005.

[16] Susnow R, Dean A, Green W, Peczak P, Broadbelt L. Rate-based construction of kinetic models for complex systems. *J Phys Chem A*. 1997;101(20):3731–3740. doi:10.1021/jp9637690.

[17] Ashcraft RW, Raman S, Green WH. Ab Initio Aqueous Thermochemistry: Application to the Oxidation of Hydroxylamine in Nitric Acid Solution. *J Phys Chem B*. 2007; 111(41):11968–11983. doi:10.1021/jp073539t.

[18] Pierotti RA. A scaled particle theory of aqueous and nonaqueous solutions. *Chem Rev*. 1976; 76(6):717–726. doi:10.1021/cr60304a002.

[19] Fredenslund A, Jones R, Prausnitz J. Group-contribution estimation of activity coefficients in nonideal liquid mixtures. *AIChE J*. 1975;21(6):1086–1099. doi:10.1002/aic.690210607.

[20] Wittig R, Lohmann J, Gmehling J. Vapor-Liquid Equilibria by UNIFAC Group Contribution. 6. Revision and Extension. *Ind Eng Chem Res*. 2003;42(1):183–188. doi:10.1021/ie020506l.

[21] Collins F, Kimball G. Diffusion-controlled reaction rates. *J Colloid Sci*. 1949;4:425–437.

Mechanism Development for Combustion of Morpholine, a Model Compound for Oxygen- and Nitrogen-Containing Fuels

Nicole Labbe, Department of Chemical Engineering, University of Massachusetts, Amherst, MA

Phillip R. Westmoreland, Chemical Engineering, University of Massachusetts Amherst, Amherst, MA

Arnas Lucassen, Department of Chemistry, Bielefeld University, Bielefeld, Germany

Patrick Oßwald, Department of Chemistry, Bielefeld University, Bielefeld, Germany

Ulf Struckmeier, Department of Chemistry, Bielefeld University, Bielefeld, Germany

Katharina Kohse-Höinghaus, Department of Chemistry, Bielefeld University, Bielefeld, Germany

Tina Kasper, Sandia National Laboratories, Livermore, CA

Nils Hansen, Sandia National Laboratories, Livermore, CA

Terrill A. Cool, School of Applied and Engineering Physics, Cornell University, Ithaca, NY

With the emergence of biofuels, oxygen- and nitrogen-containing fuels are becoming of new interest in combustion science. The fuel additive morpholine (1-oxa-4-aza-cyclohexane) is appropriate to study relevant combustion chemistry features because in one molecule, it mixes structural features of cycloalkanes, ethers, and amines.

Our collaboration at the Advanced Light Source of Lawrence Berkeley National Laboratory has measured spatial profiles of mole fraction in a premixed flat flame under low-pressure, fuel-rich (phi=1.3) conditions [1]. We were able to identify a large number of radical and stable species in the flame. Based on mechanistic insights from experiments and modeling of cyclohexane flames [2], we proposed probable reaction paths that should occur in a mechanism for morpholine combustion.

Gaussian 03 [3], ideal-gas thermochemistry from a rigid-rotor/harmonic-oscillator representation, transition-state theory, Bimolecular Quantum-RRK reaction theory [4], and the PREMIX/CHEMKIN flat-flame codes [5,6] were used to develop a morpholine model quantitatively. The Complete Basis Set method CBS-QB3 was employed to obtain thermochemical and kinetics parameters for key species and reactions in the mechanism. For many reactions, rate constants are adapted from cyclohexane kinetics, corrected for heat of reaction and reaction path degeneracy. When compared to our experimental data provided, we find good correlation between the experimental species profiles and our predictions.

Acknowledgments. This work is supported by the Division of Chemical Sciences, Geosciences, and Biosciences, the Office of Basic Energy Sciences, the U. S. Department of Energy, in part under grants DE-FG02-91ER14192 (P.R.W.) and DE-FG02-01ER15180 (T.A.C.). Sandia is a multi-program laboratory operated by Sandia Corporation, a Lockheed Martin Company, for the National Nuclear Security Administration under contract DE-AC04-94-AL85000. The Advanced Light Source is supported by the Director, Office of Science, Office of Basic Energy Sciences, Materials Sciences Division, of the U.S. Department of Energy under Contract No. DE-AC02-05CH11231 at Lawrence Berkeley National Laboratory. Support is also provided by the Deutsche Forschungsgemeinschaft under contract Ko 1363/18-3 (A.L., P.O., U.S., K.K.-H).

[1] A. Lucassen, P. Oßwald, U. Struckmeier, K. Kohse-Höinghaus, T. Kasper, N. Hansen, T.A. Cool, P.R.Westmoreland. "Species identification in a laminar premixed low-pressure flame of morpholine as a

model substance for oxygenated nitrogen-containing fuels." Proceedings of the Combustion Institute 32, 1269-1276 (2009).

[2] M.E. Law, P.R. Westmoreland, T.A. Cool, J. Wang, N. Hansen, T. Kasper. "Benzene Precursors and Formation Routes in a Stoichiometric Cyclohexane Flame." Proceedings of the Combustion Institute 31, 565-573 (2007).

[3] Gaussian 03, Revision C.02, M. J. Frisch, G. W. Trucks, H. B. Schlegel, G. E. Scuseria, M. A. Robb, J. R. Cheeseman, J. A. Montgomery, Jr., T. Vreven, K. N. Kudin, J. C. Burant, J. M. Millam, S. S. Iyengar, J. Tomasi, V. Barone, B. Mennucci, M. Cossi, G. Scalmani, N. Rega, G. A. Petersson, H. Nakatsuji, M. Hada, M. Ehara, K. Toyota, R. Fukuda, J. Hasegawa, M. Ishida, T. Nakajima, Y. Honda, O. Kitao, H. Nakai, M. Klene, X. Li, J. E. Knox, H. P. Hratchian, J. B. Cross, V. Bakken, C. Adamo, J. Jaramillo, R. Gomperts, R. E. Stratmann, O. Yazyev, A. J. Austin, R. Cammi, C. Pomelli, J. W. Ochterski, P. Y. Ayala, K. Morokuma, G. A. Voth, P. Salvador, J. J. Dannenberg, V. G. Zakrzewski, S. Dapprich, A. D. Daniels, M. C. Strain, O. Farkas, D. K. Malick, A. D. Rabuck, K. Raghavachari, J. B. Foresman, J. V. Ortiz, Q. Cui, A. G. Baboul, S. Clifford, J. Cioslowski, B. B. Stefanov, G. Liu, A. Liashenko, P. Piskorz, I. Komaromi, R. L. Martin, D. J. Fox, T. Keith, M. A. Al-Laham, C. Y. Peng, A. Nanayakkara, M. Challacombe, P. M. W. Gill, B. Johnson, W. Chen, M. W. Wong, C. Gonzalez, and J. A. Pople, Gaussian, Inc., Wallingford CT, 2004.

[4] P.R. Westmoreland, J.B. Howard, J.P. Longwell, A.M. Dean, "Prediction of Rate Constants for Combustion and Pyrolysis by Bimolecular QRRK," AIChE J., 32, 1971 (1986).

[5] R.J. Kee, J.F. Grcar, M.D. Smooke, J.A. Miller, A Fortran Program for Modeling Steady Laminar One-Dimensional Premixed Flames, Report No. SAND85-8240, Sandia National Laboratories, 1985. Modification of PREMIX version 2.5, 1991.

[6] R.J. Kee, F.M. Rupley, J. A. Miller, Chemkin-II: A Fortran Chemical Kinetics Package for the Analysis of Gas-Phase Chemical Kinetics, Report No. SAND89-8009, Sandia National Laboratories, 1989.

See more of this Session: Chemistry and Kinetics Integrated CFD Modeling
See more of this Group/Topical: Computational Molecular Science and Engineering Forum

Investigation of Reactive Center Substituent Effects On the Primary Reaction Classes During Silicon Hydride Pyrolysis: Novel Methodology for Arrhenius Parameter Estimation

Andrew J. Adamczyk, Chemical and Biological Engineering, Northwestern University, Evanston, IL

Marie-Françoise Reyniers, Chemical Engineering, Ghent University, Ghent, Belgium

Guy B. Marin, Chemical Engineering, Ghent University, Ghent, Belgium

Linda J. Broadbelt, Chemical and Biological Engineering, Northwestern University, Evanston, IL

Pyrolysis of the feed gas, typically SiH_4 or Si_2H_6, at low pressures is the standard protocol for the formation of polycrystalline silicon. Polycrystalline silicon can be formed as particles in the gas phase or controlled at a gas-solid interface to form semiconductor-grade materials through chemical vapor deposition methods. Polymerization of silicon hydrides in the gas phase causes deposits on a growing semiconductor surface forming point defects. Optimizing reactor design and process conditions plays a key role in the control of this particle formation. A clear understanding of the routes to polymerization will also allow for the control of technologies where particles are intentionally formed. These technologies create tailored nanoparticles for optoelectronic and biophotonic applications in which the size and crystallinity of the particles play integral roles.

Four major reaction classes are involved during silicon hydride pyrolysis [1]. Using high-level G3//B3LYP [2], accurate rate constants were calculated. Analysis was performed on 125 reactions for silylene-silene isomerization, Si-H bond insertion/dissociation, ring closure/opening, and hydrogen association/elimination. These reaction classes have been observed experimentally, yet rate constants cannot be measured directly for all possible reactions of silicon hydrides of relevant sizes and substituents. Thus, hydrides containing up to 10 silicon atoms, a variety of acyclic and cyclic substituents about the reactive center, and polycyclic nature were explored.

Benson's group additivity model was extended to predict the Arrhenius parameters, the pre-exponential factor (A) and activation energy (E_a), for the major reaction classes during silicon hydride pyrolysis. The transition state group additivity (TSGA) model, previously applied to the pyrolysis of hydrocarbons [3], predicts rate constants with superior accuracy compared to the currently used Evans-Polanyi relationship for E_a and a representative value of A for each reaction class. The TSGA method possesses several advantages that should interest kineticists seeking parameters for the primary reaction families in silicon hydride formation: (1) allows one to gain insight into the effect of structural changes on the reactive center during reaction, (2) implements more explicit values of A for each reaction class, and (3) circumvents the need to calculate accurate enthalpies and Gibb's free energies of reaction as required for the Evans-Polanyi correlation and subsequent calculation of the reverse rate constant, respectively. The knowledge of the number and the type of substituents on the reactive center for a given reaction class is sufficient to allow accurate kinetic parameter estimation.

Keywords: kinetics, nanoparticles, intermediates, G3//B3LYP, group additivity

---

[1] Wong, H. W.; Li, X. G.; Swihart, M. T.; Broadbelt, L. J. *Journal of Physical Chemistry A* **108** 10122 (2004)

[2] Baboul, A. G., Curtiss, L. A., Redfern, P. C., and Raghavachari, K. *J Chem Physics* **110** 7650 (1999)

[3] Sabbe, M. K., Reyniers, M. F., Van Speybroeck, V., Waroquier, M., Marin, G. B. *Chemphyschem* **9** 124 (2008)

See more of this Session: Chemistry and Kinetics Integrated CFD Modeling
See more of this Group/Topical: Computational Molecular Science and Engineering Forum

Hydrogen Production From Carbohydrates by Reactive Dynamics with the Reactive Force Field

Luzheng Zhang, Petroleum Recovery Research Center, New Mexico Institute of Mining and Technology, Socorro, NM

Robert Lee, Petroleum Recovery Research Center, New Mexico Institute of Mining and Technology, Socorro, NM

Production of hydrogen from renewable carbohydrates is considered an effective approach to mitigating the environmental problems caused by CO2 emissions from fossil fuels. However, recent studies on hydrogen production from reforming liquid phase carbohydrates were performed relatively high pressures and temperatures with a high loading of novel metal particles. On the other hand, the reactivity of catalysts and their selectivity have room to improve and there is a lack of understanding on the interactions between catalyst particles and their support materials, particularly on the interfacial catalytic reactions on transition metal surfaces (e.g., Pt).

In this work, chemical reactions for hydrogen production from methanol and glucose aqueous solutions were studied by using the reactive dynamics (RD) with reactive force field (ReaxFF). We carried out a number of RD simulations with ReaxFF for reactive systems in the bulk and in the presence of Pt catalyst surfaces over a wide range of temperature and pressure. Preliminary results showed that in the bulk reactive systems, hydrogen production rate increases with reaction temperature and that pressure has less effect on hydrogen production rate than temperature. In the presence of Pt catalysts, the temperature at which hydrogen is produced is decreased and Pt[100] surface seems to be more active than Pt[111] surface for hydrogen production from carbohydrates. In addition, we compared some of simulation results with hydrogen production from methanol-water aqueous solution on Pt/TiO2.

See more of this Session: Chemistry and Kinetics Integrated CFD Modeling
See more of this Group/Topical: Computational Molecular Science and Engineering Forum

Integration of EFPS and PROVISION for Hybrid Cracking in Modern Cracking Furnace Design

Harry Wang, Bruce Evans

Technip USA, Inc., 555 West Arrow Highway, Claremont, CA 91711-4805, U.S.A.

Abstract

Cracking furnaces are the heart of the ethylene plant. Ethylene producers purchase various feedstocks ranging from ethane, propane, EP, NGL's, naphtha and up to heavy condensates and vacuum gas oil (VGO). However, in a modern feed flexible ethylene plant, the quantity of each feedstock received varies. In co-cracking or blended cracking, each feedstock often cannot be cracked at its optimal conversion or severity. Economically, it is also very important to minimize number of furnaces in the design of an ethylene plant.

Therefore, hybrid cracking (two or more feeds cracked in one single cell or twin cell furnace) with partial decoking capability (cracking in one cell while decoking in another cell in twin cell furnace) becomes important in modern furnace design for achieving feedstock flexibility while maintaining furnace availability. However, a significant challenge in the design of hybrid cracking furnaces is how to properly simulate the hybrid furnace performance.

This paper presents a novel cracking furnace model developed by Technip called **SPLIT-MIXING** to simulate hybrid cracking using an integrated EFPS model and PROVISION software. This program conveniently runs in PROVISION environment.

Computational Fluid Dynamics (CFD) is also applied to demonstrate furnace flue gas flow distribution in the transition region from firebox to convection section and in the lower banks of the convection section where flue gas is relatively hot.

This novel model can be used and extended for simulating hybrid cracking with two gas feedstocks, one gas plus one liquid feed, or two different liquid feedstocks.

The furnace can be single cell or twin cell. For a twin cell furnace, one cell can be on gas or liquid feed cracking while another cell is on steam air decoking.

Introduction

In a modern feed flexible cracker design, one of the challenges in the engineering phase is to properly design flexible feed cracking furnaces. With the advancement of ethylene furnace technologies, hybrid cracking (or called segregated cracking or split cracking) furnaces are often taken into consideration by engineering licensors and ethylene producers in order to optimize furnace configuration, minimize number of furnaces and achieve optimal furnace cracking conversion for gas cracking or optimal cracking severity for liquid cracking. However, the challenge in the design of hybrid cracking furnaces is how to correctly simulate the furnace performance due to the following reasons:

- Two or more feedstocks as well as mixed feed (dilution steam plus feed) enter different passes of process banks in the convection section and different groups of radiant coils in the firebox. However, high pressure steam super heater coils and boiler feed water coils are still common in the convection section. Accurate prediction of process and flue gas temperature profiles is a challenge.

- Fired duty in each cell of a twin cell furnace during on line steam-air decoking in one cell and cracking in another cell can be significantly different. Challenges in the design are the prediction of flue gas mixing in the transition zone from the radiant section of each cell to the common convection section, associated impact on the firebox, and prediction of the performance of the coils and passes of the convection section.

A novel cracking furnace model called **SPLIT- MIXING** was developed by Technip to accurately simulate hybrid cracking using an integrated EFPS model and PROVISION software. This program conveniently runs in PROVISION environment.

EFPS

EFPS (**E**thylene **F**urnace **P**rogram **S**et) is Technip's proprietary ethylene furnace yield prediction and cracking furnace simulation program, which is used by nearly 80% of worldwide Ethylene Producers for feedstock selection, case studies, trouble shooting and optimal furnace operation. EFPS program is also Technip's standard program to design, develop and engineer ethylene cracking furnaces.

EFPS consists of four sub models:

- SPYRO (**S**team **Pyro**lysis) - Rigorous Radiant Coil Simulation Program. More than 3500 reactions are included in SPYRO 6 model of EFPS program. This model consists of 128 molecular species and 20 chain propagating radicals and up to C_{42} hydrocarbons. SPYRO model is a result of more than 35 years of research and development. SPYRO model can also be used to predict radiant coil run length.
- FIREBOX - Rigorous Firebox Simulation Program. This model was developed based on Sarafim and Hottel's firebox model by the use of zoning method for the calculation of radiative heat transfer. By combination of SPYRO with FIREBOX, this model is used to accurately predict heat flux profiles, radiant tube metal temperature as well as flue gas and refractory temperature profiles.

- CONVEC - Rigorous Convection Section Simulation Program. This model is used to accurately calculate heat transfer of each bank, predict both process and flue gas side temperature and pressure profiles as well as tube and fin metal temperatures in each row of convection banks.
- TES – Rigorous Transferline Exchanger Simulation Program. This program is used to predict heat transfer for the transfer line exchangers by the linkage with steam drum through risers and downcomers. The transferline exchangers can be boilers or boiler feed water pre-heaters or feed pre-heaters. TES can also be used to predict TLE run length.

Figure 1 is the EFPS program structure which shows the interaction between these four sub models. EFPS links these four models and calculates furnace overall performance.

EFPS/PROVISION Linkage

PROVISION (PROII) is well known process engineering software which is also widely applied in the petrochemical industry for process design, simulation and optimization.

EFPS is used to simulate SPYRO, FIREBOX, CONVEC and TES simultaneously. However, EFPS itself can be used to simulate one feedstock but not multiple feedstocks in the furnace. This is because SPYRO in EFPS can specify one or more than one type of feeds for co-cracking (blended) application but not two or more than two feeds in one SPYRO file for segregated cracking.

In order to properly simulate hybrid furnace performance with more than one feed in either single cell or twin cell furnaces, Technip developed a **SPLIT-MIXING** model to link EFPS with PROVISION to solve two or more than two SPYRO files simultaneously. The program is conveniently executed in PROVISION environment.

The convection section is in common although different feeds can be cracked in separate radiant coils of firebox. Therefore, on the flue gas side, when flue gas enters process bank of the convection section, the flue gas is assumed to be split into different streams before entering the process bank. The fraction of **SPLITTER** is based on how many passes of process bank in the convection section are in one feed and how many passes of process bank in the convection section are in another feed (assuming two feedstocks only). Whenever flue gas enters very high pressure steam banks or very high pressure economizer bank, different streams of flue gas coming out from the process bank are assumed to be combined in a **MIXER** before entering steam banks or economizer bank. The steam banks, economizer bank and steam drum are in common.

Since **SPLITTER** and **MIXER** can be conveniently defined in the PROVISION environment, this hybrid cracking model is called a **SPLIT- MIXING** model.

EFPS, a user defined UNIT is created in PROVISION environment. PROVISION calls UNIT "FIREBOX_SPYRO" in EFPS program at each calculation sequence to obtain flue gas and bridge wall conditions. UNIT "Heat Exchanger" (simple HX) in PROVISION is used to match performance of CONVEC and TES from EFPS program. Heat transfer coefficients and logarithmic delta T calculated from CONVEC and TES are used as an input to the UNIT "Heat Exchanger" in PROVISION.

An automatic steam balance is also developed in **SPLIT- MIXING** model.

Figure 2 is a typical sketch to present an EFPS-PROVISION linkage to simulate hybrid furnace with two feedstocks.

EFPS-PROVISION integration program for hybrid cracking with two feedstocks is divided into the following eight (8) sections:

- Feed A preheat and DS preheat as well as mixed feed A plus steam preheat sections.
- Feed B preheat and DS preheat as well as mixed feed B plus steam preheat sections.
- Firebox/SPYRO with Feed A.
- Firebox/SPYRO with Feed B.
- TLEs for cracked gas heat recovery from mixed feed A cracking.
- TLEs for cracked gas heat recovery from mixed feed B cracking.
- Convection section with all the banks including individual process banks as well as common steam and BFW banks.
- Steam balance to calculate BFW/export steam flow / desuperheater water / steam drum blowdown.

This model can be extended to simulate:

- Cracking with more than two feedstocks in single cell or twin cell,
- Cracking with one feed stock in one cell and steam air decoking in another cell for twin cell furnaces. Decoke effluent can either go to decoke cyclone or be recycled back to the bottom of firebox for complete combustion.

Case Studies

Three case studies are presented in this paper:

- Gas-Gas (**G-G**) hybrid cracking for single cell furnace with ethane/propane segregated cracking
- Gas-Liquid (**G-L**) hybrid cracking for twin cell furnaces with propane/naphtha segregated cracking.
- Gas-Decoke (**G-D**) hybrid cracking for twin cell furnaces with partial steam air decoking in one cell and ethane/propane mixed (EP) feed cracking in another cell.

1. **G-G** hybrid cracking for single cell furnace with ethane/propane segregated cracking capabilities

In this case, ethane and propane feeds are segregated cracked in SMK furnace coils.

SMK is a trademark of Technip developed radiant coils for gas feed cracking.

Half of the radiant coils are supplied with ethane feed and half of the radiant coils are supplied with propane feed. There are eight passes of convection section with four passes supplied with ethane feed and four passes supplied with propane feed. There are also two or more firing zones in each half of the firebox.

A hybrid cracking single cell model has been developed. Figure 3 is the EFPS-PROVISION model developed in PROVISION environment.

In this furnace design with zone firing and pass balancing capabilities, ethane feed flow is fixed and propane feed flow is floating until furnace duty contributed from ethane feed is the same as furnace duty contributed from propane feed to achieve zone firing balance in the firebox. A controller is defined in the PROVISION to let propane feed flow float until firing rates from both sides are the same.

Heat transfer results from CONVEC program in EFPS from both process sides with different feeds as well as common steam and economizer banks are transferred into various simulation UNITs as heat exchangers (Simple HX) in the PROVISION model to simulation convection section performance.

Similarly, heat transfer results from TES program in EFPS are also transferred into various simulation UNITs as heat exchangers (Simple HX) in the PROVISION model to simulate transfer line exchanger performance.

Overall steam balance is also developed in this PROVISION model to calculate furnace export steam and boiler feed water consumption.

Figure 4 is the result of G-G hybrid single cell furnace performance from the EFPS-PROVISION simulation.

2. G-L hybrid cracking for twin cell furnaces with propane/naphtha segregated cracking in the radiant coils.

Propane and naphtha feeds are segregated cracked in a twin cell furnace with GK-6 radiant coils in each cell.

GK-6 is a trademark of Technip developed radiant coils for flexible feed cracking or liquid feed cracking.

In this particular case, one and a half cells are cracked with propane feed and half of another cell is cracked with naphtha feed. There are eight passes of convection section. Therefore, six passes of the feed bank, dilution steam bank, feed plus steam mixed bank are supplied with propane feed as well as its own dilution steam and two passes of the feed bank, dilution steam bank, feed plus steam mixed bank are supplied with naphtha feed as well as its own dilution steam. There are also two or more firing zones with each cell of the firebox.

A hybrid cracking twin cell model has been developed. Figure 5 is the EFPS-PROVISION model developed in PROVISION environment.

In this model with zone firing and pass balancing capabilities, propane feed flow is fixed and naphtha feed flow is floating until furnace duty with propane feed in one cell is the same as furnace duty with propane/naphtha segregated feed in another cell to achieve firing balance between cells. It is important to balance the firing duty generated from each cell so that flue gas flow from each of the two cells are in balance. This will prevent from flue gas mal-distribution particularly in the transition area from each cell of the firebox to the lower part of the convection section. A controller is defined in the EFPS – PROVISION program to float one feed flow until firing rates from both cells are equal.

Heat transfer results from CONVEC program in EFPS from both process sides with different feeds as well as common steam and economizer banks are transferred into various simulation UNITs as heat exchangers (Simple HX) in the PROVISION model to simulation convection section performance.

Similarly, heat transfer results from TES program are also transferred into various simulation UNITs as heat exchangers (Simple HX) in the PROVISION model to simulate transfer line exchanger performance.

Overall steam balance is also developed in this PROVISION model to calculate furnace export steam and boiler feed water consumption.

A special Wegstein acceleration method is applied for fast convergence.

Figure 6 is the result of G-L hybrid twin cell cracking furnace performance from the EFPS-PROVISION simulation.

3. Hybrid cracking in twin cell furnaces with EP (ethane/propane mixed feed) cracking in one cell and on line air-steam decoking recycled back to firebox in another cell.

In this case, EP feed is cracked in one cell with GK-6 radiant coils at high conversion of ethane and on line steam-air decoking in another cell also with GK-6 radiant coils.

There are also eight passes of convection section process banks. Therefore, four passes of the feed bank, dilution steam bank, feed plus steam mixed bank are supplied with ethane feed, dilution steam and ethane feed plus steam and four passes of the feed bank, dilution steam bank, feed plus steam mixed bank are supplied with air, steam and air plus steam feed. There are also two or more firing zones in each cell of the firebox.

A hybrid cracking twin cell model with partial decoking capabilities has been developed. Figure 7 is the EFPS-PROVISION model developed in PROVISION environment.

In this model, since fired duty from EP cracking is much higher than decoke cell, a special distribution wall located at the transition area of arch is designed to balance the flue gas flow distribution and minimize flue gas recirculation. Computational Fluid Dynamics (CFD) model is developed to confirm the flue gas distribution in the arch area and convection section area based on the design of this special distribution wall.

Figure 8, 9, 10 and 11 are the CFD results to demonstrate a free standing wall design to straighten the flue gas flow pass and minimize flue gas recirculation or mal-distribution in the transition area from fireboxes of each cell to the common shared convection section.

In general, the bridge wall temperature from cracking EP cell is higher than the bridge wall temperature from decoke side. In addition, flue gas flow from EP cracking cell is also generally quite a bit higher than the flue gas flow from steam air decoke cell. Upon flue gas mixing in the arch area, if special design is not done in the lower part of the convection section, crossover temperature outlet of decoke air plus steam bank will be too high and may even exceed the design temperature of crossover piping which typically is SS-304 material.

A special temperature control scheme is developed by Technip to prevent from over heating the furnace crossover pipings in the cell with steam air decoking.

Heat transfer results from CONVEC program in EFPS from both process sides with different feeds as well as common steam and economizer banks are transferred into various simulation UNITs as heat exchangers (Simple HX) in the PROVISION model to simulate convection section performance.

Similarly, heat transfer results from TES program are also transferred into various simulation UNITs as heat exchangers (Simple HX) in the PROVISION model to simulate transfer line exchanger performance.

Overall steam balance is also developed in this PROVISION model to calculate furnace export steam and boiler feed water consumption.

A special Wegstein acceleration method is applied for fast convergence.

Figure 12 is the results of **G-D** hybrid twin cell cracking furnace performance from EFPS-PROVISION simulation.

Summaries and Conclusions

This novel hybrid cracking **SPLIT- MIXING** model can be used and extended for handling various hybrid cracking cases with two or more feeds, one gas feed plus one liquid feed or two or more different liquid feeds.

The furnace can be single cell or twin cell types. For a twin cell furnace, one cell can be on gas or liquid feed cracking while another cell is on steam air decoking. Decoke effluent can be routed to cyclone or recycled back to firebox. Radiant coils can also be different in each cell with different feedstocks.

This model has been successfully applied in the design and engineering of Technip designed ethylene crackers.

FIGURE 1: EFPS - PROGRAM STRUCTURE

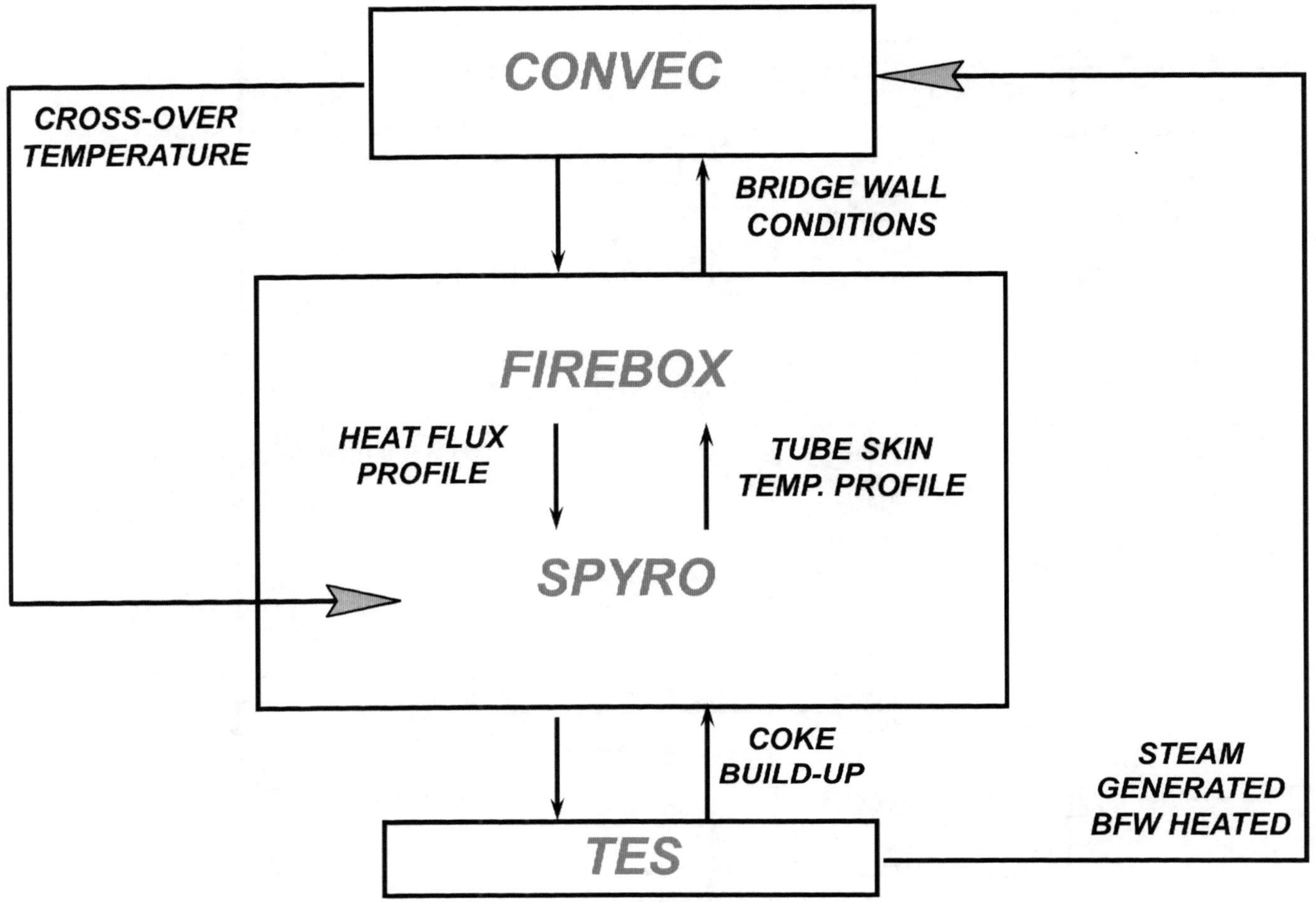

Technip

FIGURE 2: SPLIT-MIXING HYBRID CRACKING MODEL USING EFPS-PROVISION LINKAGE

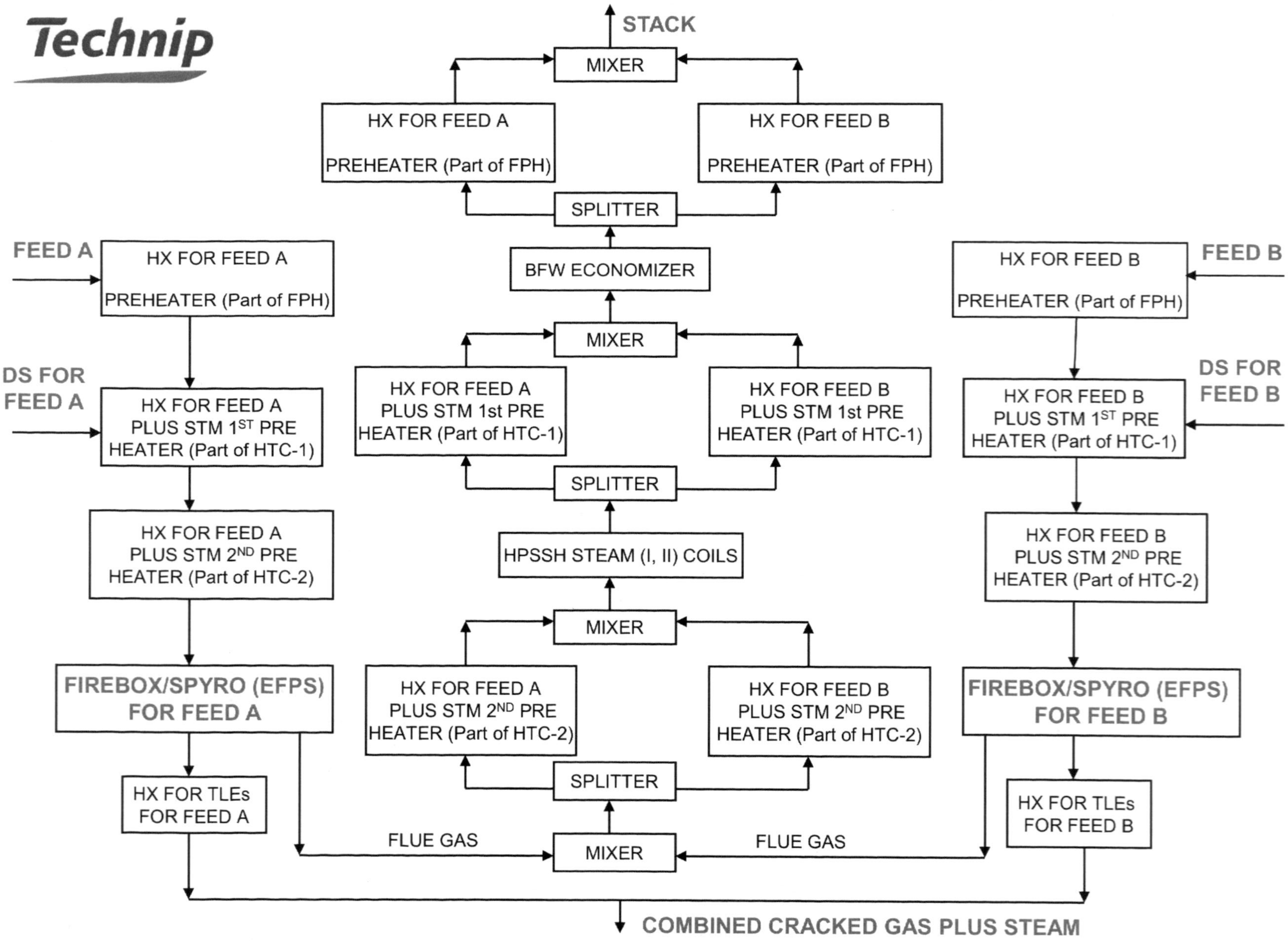

Technip

Figure 3: G-G Hybrid Cracking PROVISION-EFPS Model in Single Cell SMK Furnace

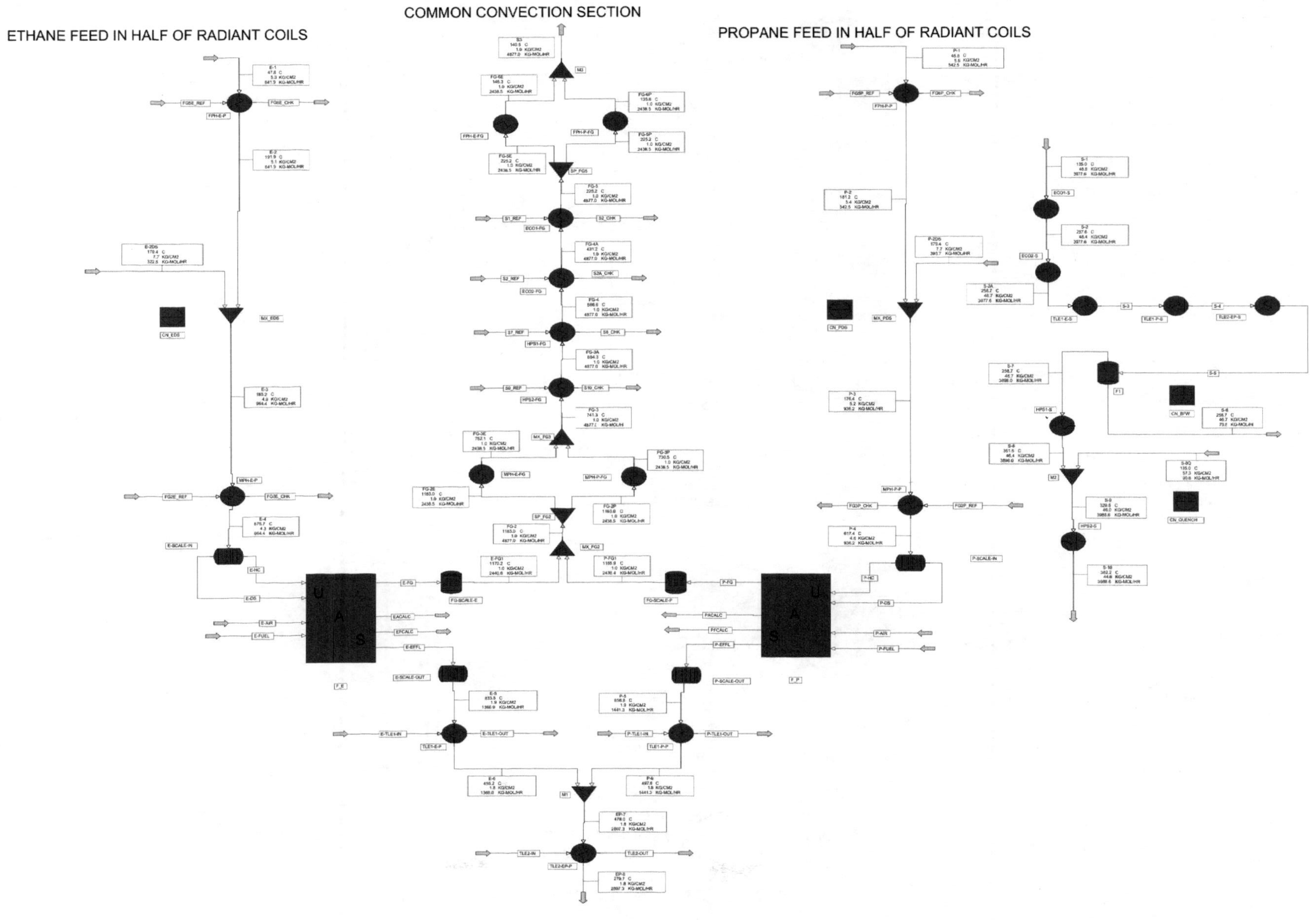

FIGURE 4: FURNACE PERFORMANCE FOR GAS - GAS HYBRID SINGLE CELL CRACKING

Technip

SMK HEATER, SINGLE CELL DESIGN
HYBRID CRACKING: 1/2 FURNACE ETHANE FEED, 1/2 FURNACE PROPANE FEED
CRACKING CONVERSION OF ETHANE: 67% CONVERSION OF ETHANE
CRACKING CONVERSION OF PROPANE: 92% CONVERSION OF PROPANE

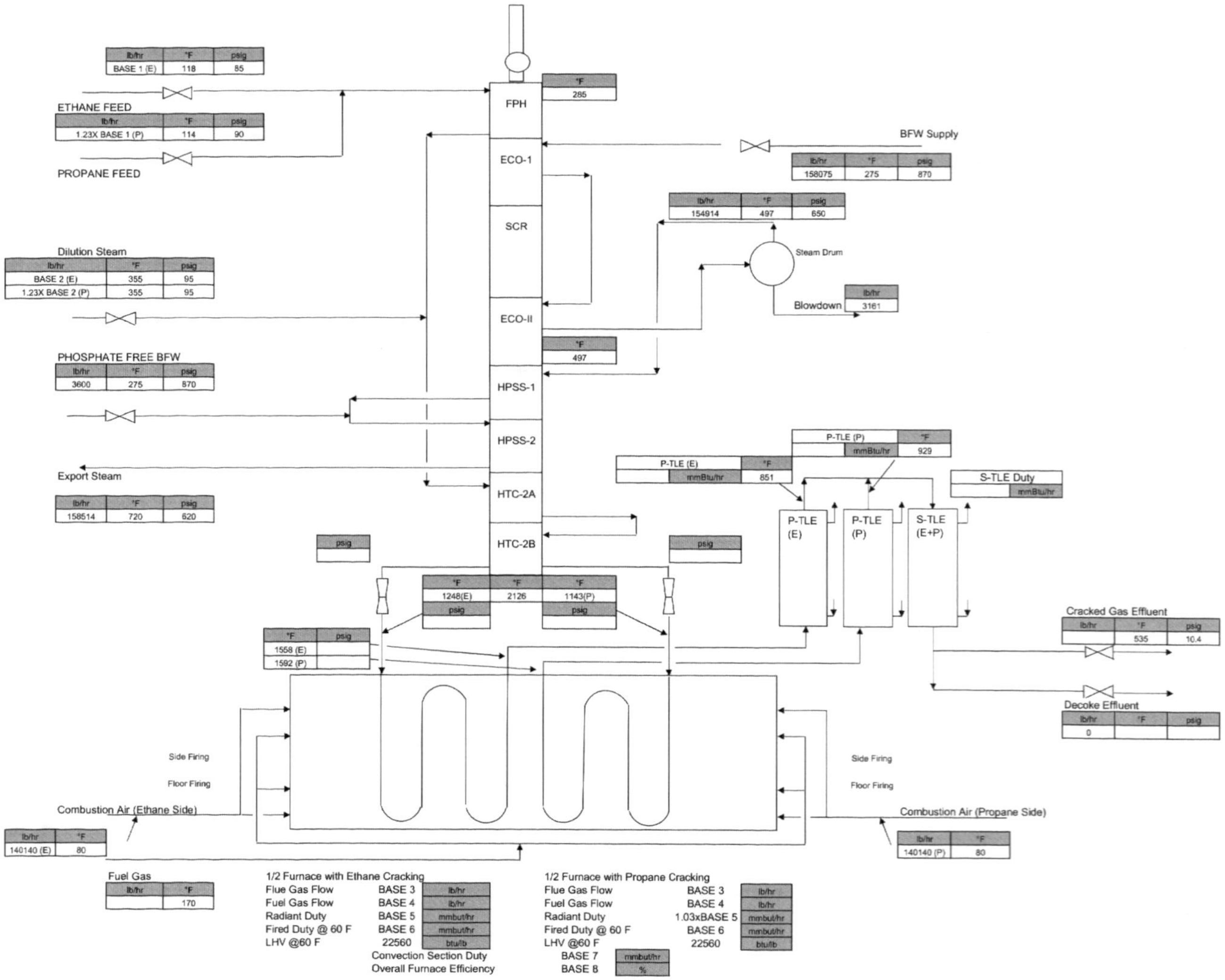

Figure 5: G-L Hybrid Cracking in Twin Cell GK-6 Furnace

COMMON CONVECTION SECTION

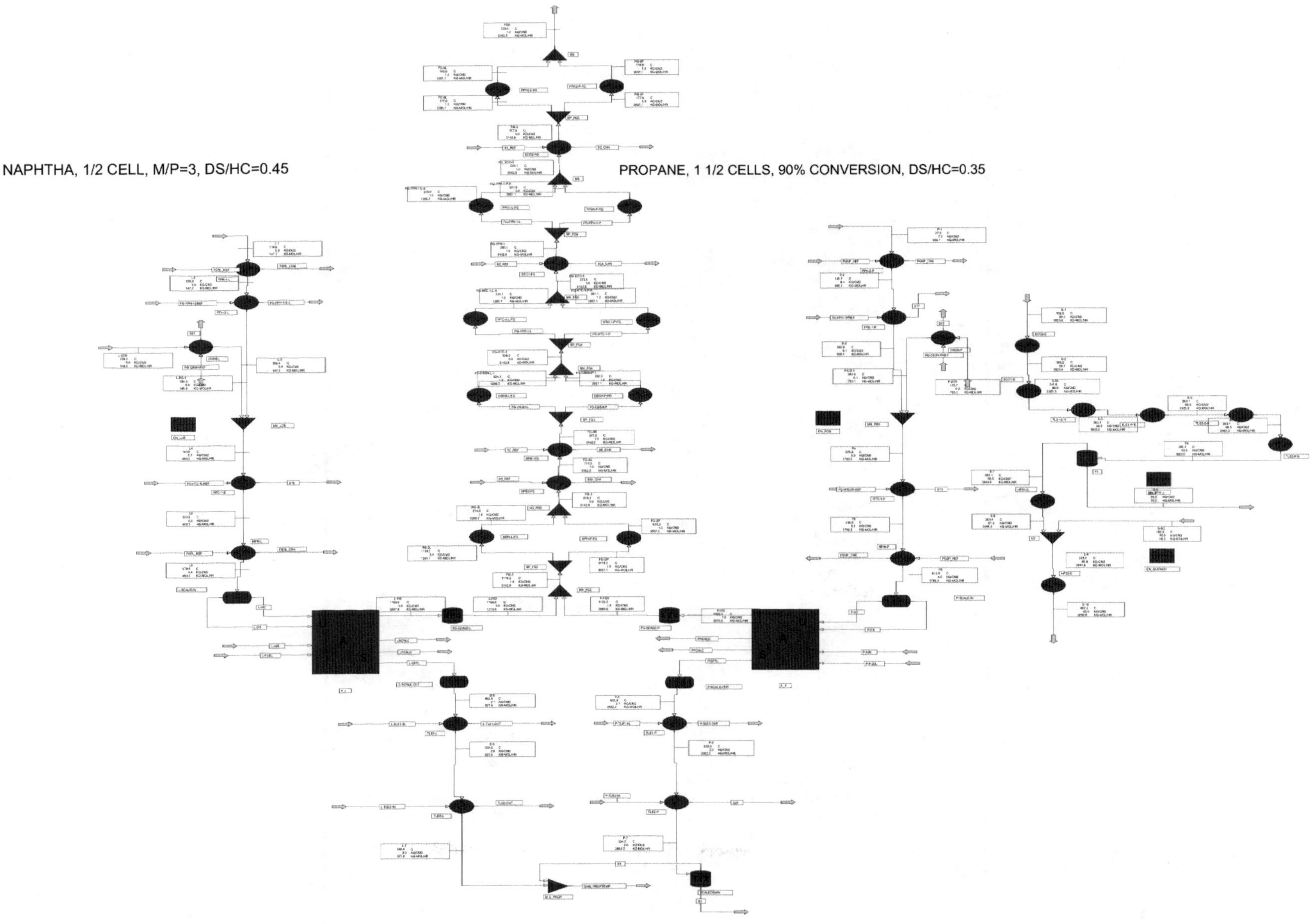

FIGURE 6: FURNACE PERFORMANCE FOR GAS - LIQUID HYBRID TWIN CELL CRACKING

Technip

GK-6 HEATER, TWIN CELL DESIGN
HYBRID CRACKING: 1 1/2 CELLS PROPANE FEED, 1/2 CELL NAPHTHA FEED
PROPANE CRACKING CONVERSION: 90% CONVERSION OF PROPANE
NAPHTHA CRACKING SEVERISTY: METHANE/PROPANE = 3 (MOLAR RATIO)

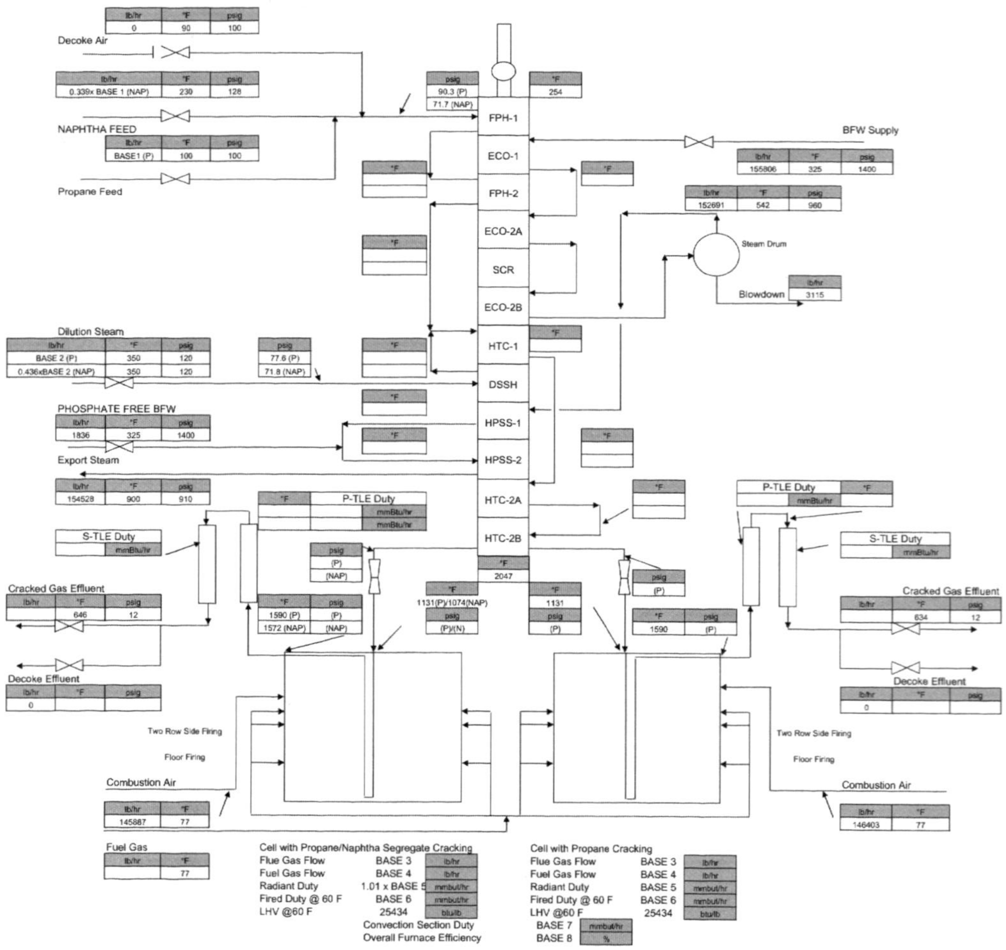

Figure 7: Gas_Decoke Hybrid Cracking In Twin Cell Furnace

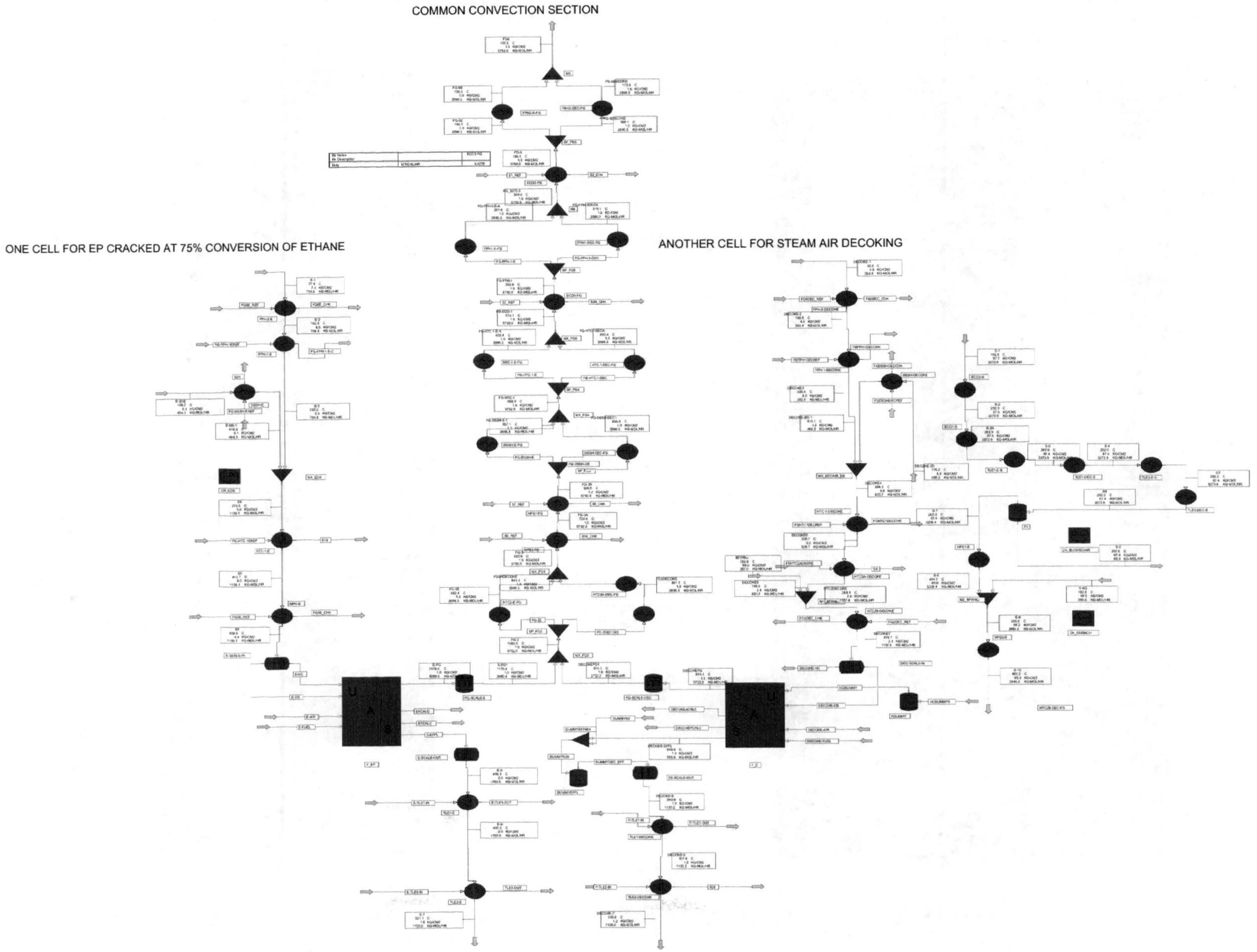

Figure 8: Model Geometry (With/without Distribution Wall)

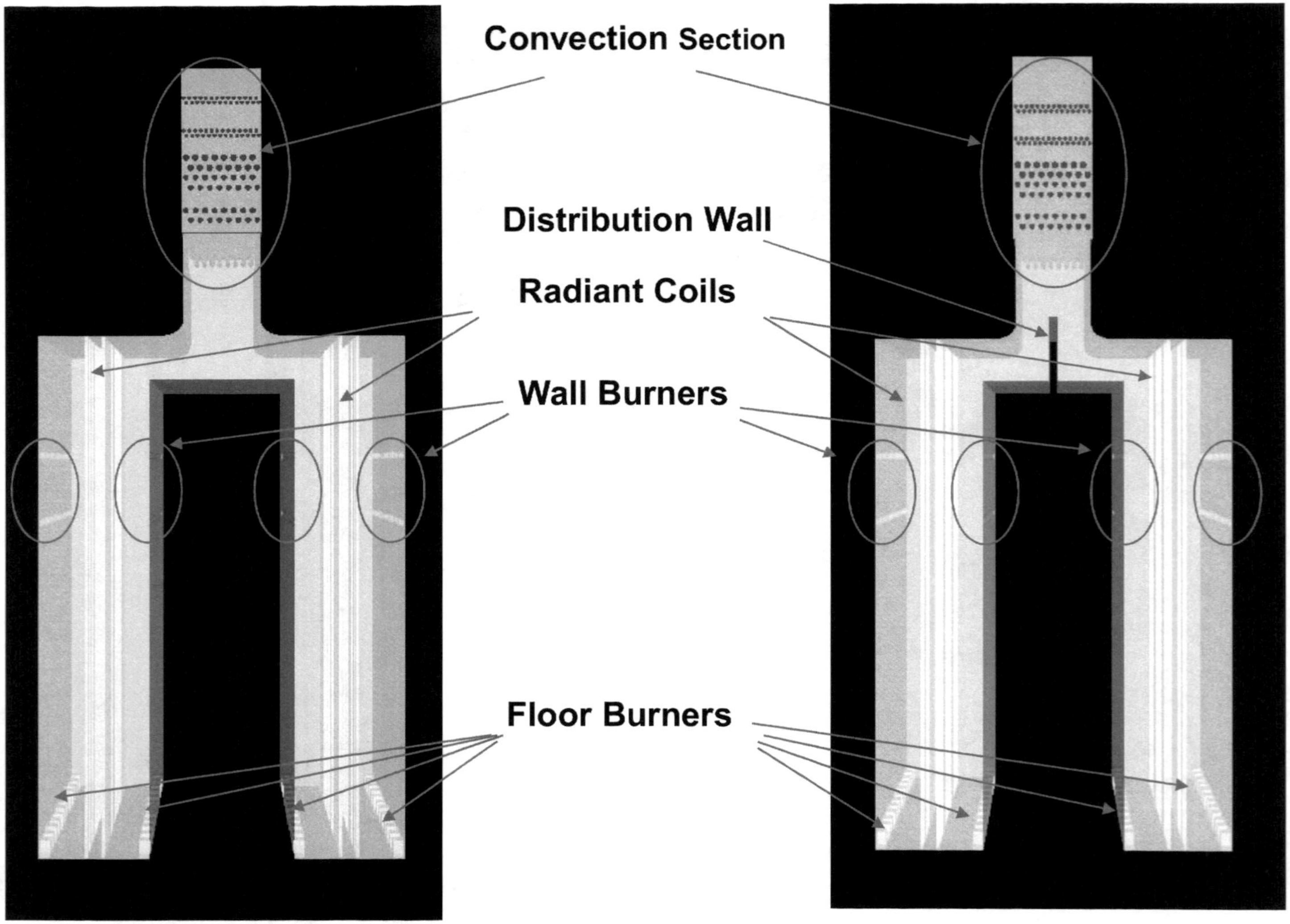

Technip

Figure 9: Velocity contours through convection section (ft/s)

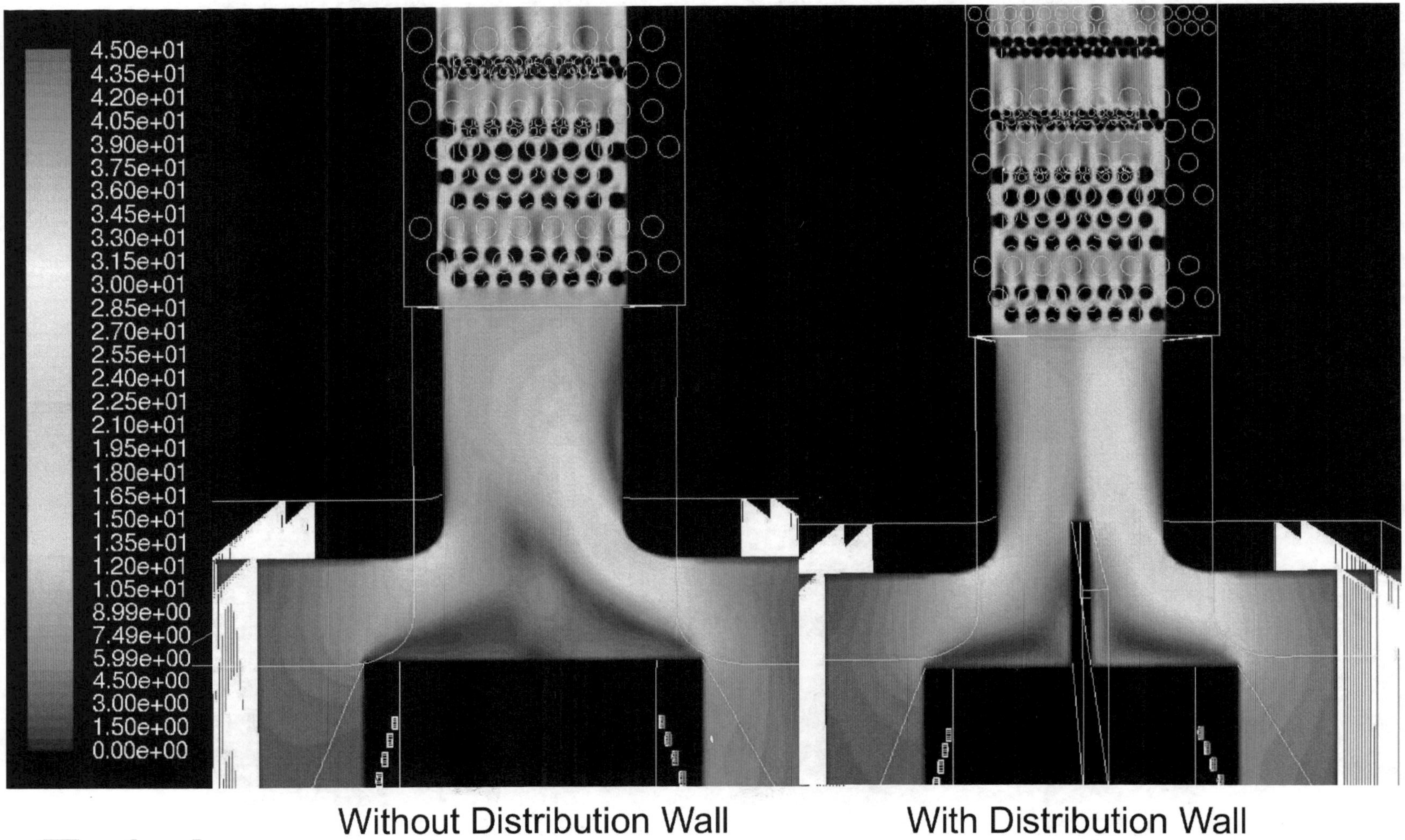

Figure 10: Temperature contours (C) at the center of furnace

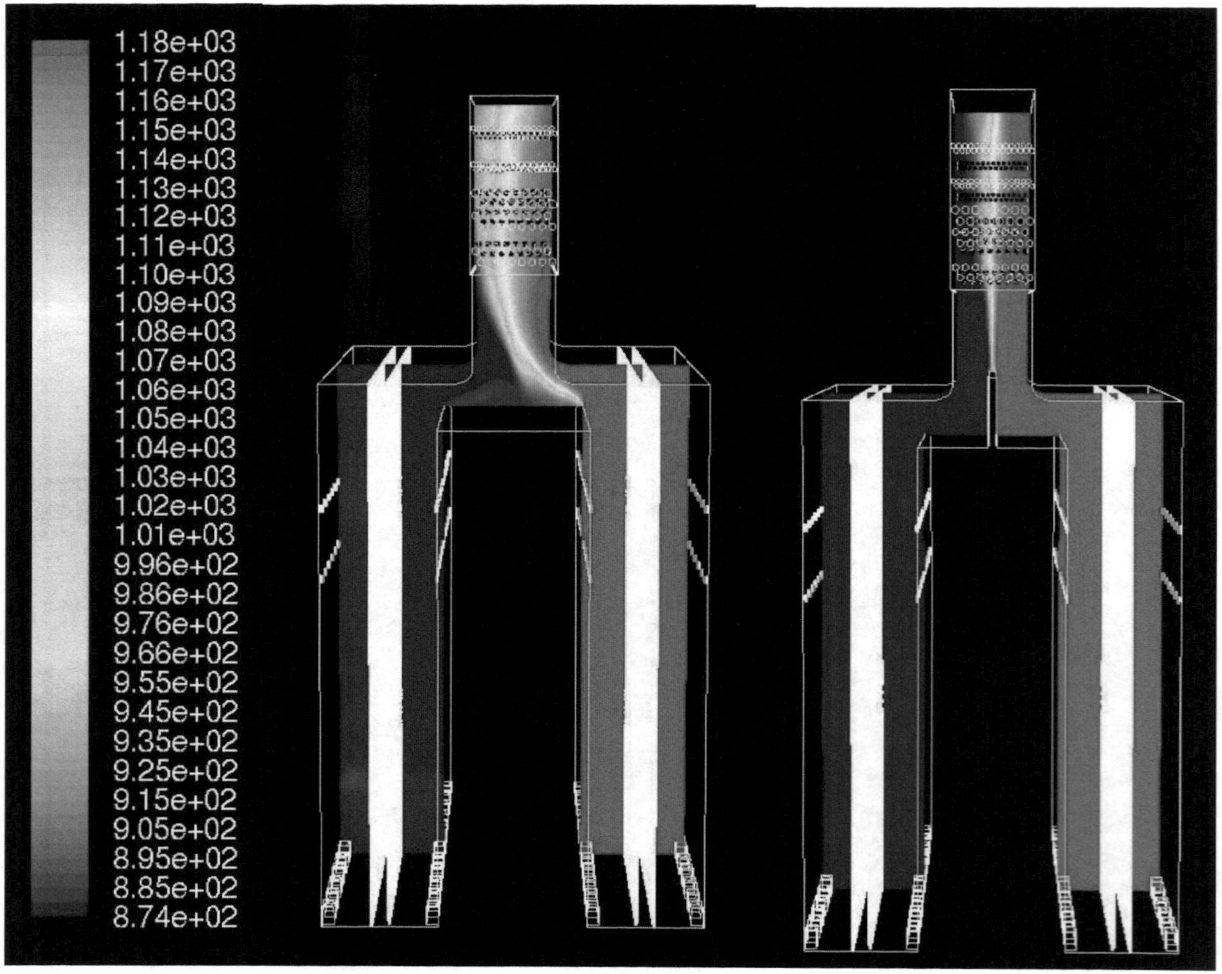

Technip

Without Distribution Wall With Distribution Wall

Figure 11: Temperature contours (C) before first coil

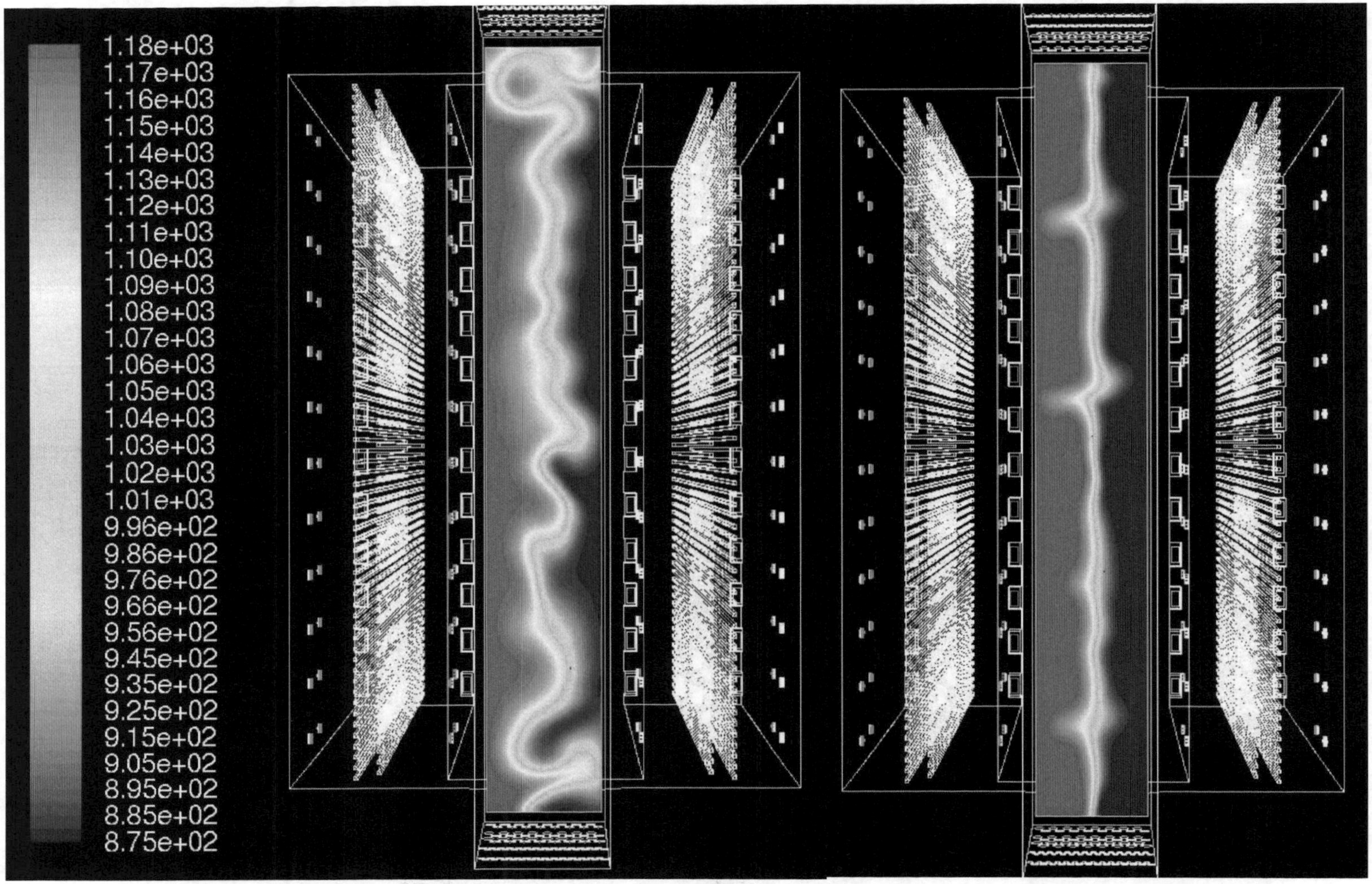

Without Distribution Wall   With Distribution Wall

Technip

FIGURE 12: FURNACE PERFORMANCE FOR GAS - DECOKE HYBRID TWIN CELL CRACKING

Technip

GK-6 HEATER, TWIN CELL DESIGN
DECOKE: ONE CELL EP FEED CRACKING, ANOTHER CELL STEAM AIR DECOKING
EP CELL WITH 70 WT% ETHANE, 75% CONVERSION OF ETHANE

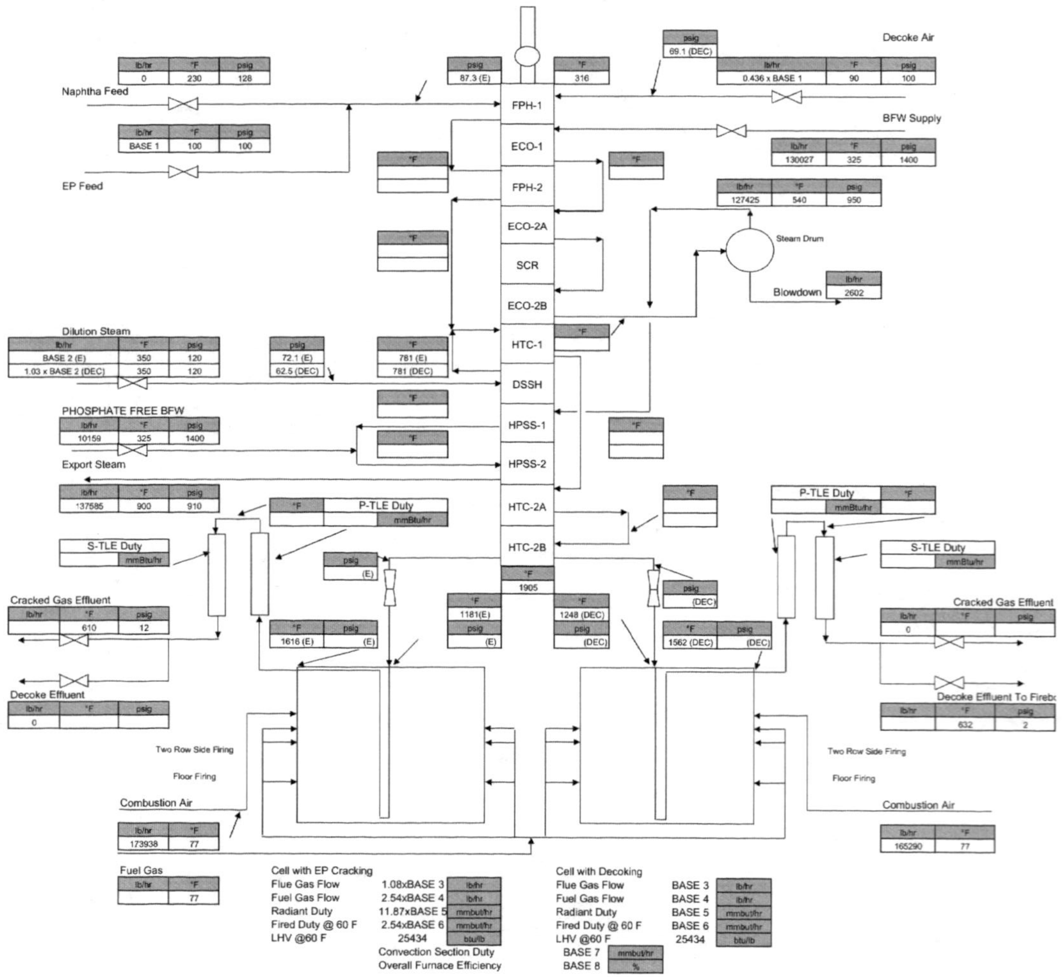

Theoretical Investigation of the Asymmetric Alkylation of Benzaldehyde in Presence of Organozinc Reagents

Klaus B. Stark, Scienomics SARL, Houston, TX

Xenophon Krokidis, Scienomics SARL, Houston, TX

Andreas Bick, Scienomics SARL, Houston, TX

In this paper the asymmetric alkylation of benzaldehyde is revisited by means of semi-empirical and density functional theory based transition state modeling. The reaction proceeds in presence of a dialkylzinc and a b-aminoalkohol which form an amino-coordinated zinc-alkoxide with the benzaldehyde. Then a second equivalent of diethylzinc is added to form a supermolecular transition state for the reaction. Several chiral aminoalkohols have been shown to give a high enantiomeric excesses and the computed stabilities of the supermolecular transition state structures are compared with the available experimental data giving an overall good agreement with the experimental trends.

See more of this Session: Chemistry and Kinetics Integrated CFD Modeling
See more of this Group/Topical: Computational Molecular Science and Engineering Forum

Surface Atomic Distribution and Water/Atomic Oxygen Adsorption On Pt-Co Alloys

Pussana Hirunsit, Chemical Engineering, Texas A&M University, College Station, TX

Perla B. Balbuena, Chemical Engineering, Texas A&M University, College Station, TX

Pt-based alloy surfaces are used to catalyze the molecular oxygen and its subsequent reduction to water on fuel cell electrodes. Due to surface segregation phenomena, alloy surface composition may be very different from the overall composition. Also, the adsorption of intermediates or products species on the surface may modify the properties of both surfaces and species. Here we address the problem of determining surface atomic distribution using density functional theory (DFT) calculations on slabs of PtCo and Pt_3Co overall compositions, as well as a water molecule adsorption on PtCo(111) and Pt-skin structures. Pt-rich surfaces are energetically favored under vacuum in the PtCo and Pt_3Co alloys. The adsorption trend on the studied structures agrees with the d band model, with stronger adsorption at higher surface Co composition. The bonding between water and surfaces of PtCo and Pt-skin monolayers are investigated in term of energetic and electron orbitals. We also found that water adsorption on Pt-skin shows larger effect on atomic oxygen binding energy than that on PtCo. On both surfaces, the surface reconstruction occurs at high oxygen coverage (~0.5 ML) but the presence of water at low oxygen coverage of 0.125 ML can present a comparable magnitude of surface reconstruction. The strength of O adsorption, as well as surface reconstruction effects due to the adsorbate are strongly influenced by the presence of the oxyphilic transition metal on the surface or subsurface.

See more of this Session: First-Principles Simulation of Condensed Phases: Surfaces
See more of this Group/Topical: Computational Molecular Science and Engineering Forum

Chemical Erosion of Silica Nitride Ceramics

Jan Andzelm, U. S. Army Research Laboratory, Aberdeen Proving Ground, MD

Jennifer Synowczynski, Weapons and Materials Research Directorate, U.S. Army Research Laboratory, Aberdeen Proving Grounds, MD

Jeffrey Swab, U. S. Army Research Laboratory, Aberdeen Proving Ground, MD

The high temperature mechanical and fracture properties of bulk ceramics such as Si3N4 and SiC are dominated by the strength and chemical stability of the thin intergranular glassy film (IGF) that forms during the liquid phase sintering process. This IGF interface is susceptible to chemical erosion, particularly at elevated temperature and pressure. We are interested in understanding the mechanism of chemical erosion of the silica-nitride material exposed to gases such as carbon monoxide or water at elevated temperature. It has been previously demonstrated that the weak IGF interface can be strengthened by using sintering additives such as rare-earth oxides which segregate to the IGF and induce local ordering. In this study, we performed solid-state, density functional (DFT) simulations of chemisorption, surface and in-bulk diffusion for various models of the IGF interface. The impact of rare-earth oxides on chemisorption and diffusive processes will also be discussed.

See more of this Session: First-Principles Simulation of Condensed Phases: Surfaces
See more of this Group/Topical: Computational Molecular Science and Engineering Forum

Density Functional Theory Studies of Doping in Both Bulk- and Surface-State Titania

Run Long, Chemical Engineering, University College Dublin, Dublin, MT, Ireland

Niall J. English, Chemical and Bioprocess Engineering, University College Dublin, Dublin, MT, Ireland

The structural, energetic and electronic properties of substitutional Ge, N/W, Bi/S (co)doped bulk anatase-titania systems have been investigated, along with those of a range of adsorptive and substitutional S- and P-doped rutile-titania (110) surfaces, by first-principles density functional theory (DFT) calculations. The stability of the various doping configurations was compared on the basis of their calculated formation/adsorption energies, and preferred dopant states were identified and characterised in terms of structural features. In many cases, red-shifts in optical absorption edge and reductions in photon transition energy were observed, along with synergistic effects between electronic states introduced by co-dopants. These observations have been used to rationalise various recent experimental observations under these dopant conditions.

See more of this Session: First-Principles Simulation of Condensed Phases: Surfaces
See more of this Group/Topical: Computational Molecular Science and Engineering Forum

Study of Cobalt Surface and Crystal Defect Behavior Using a Reactive Force Field (ReaxFF)

Matthew R. LaBrosse, Chemical Engineering, University of Pittsburgh, Pittsburgh, PA

Adri C.T. van Duin, Department of Mechanical and Nuclear Engineering, Pennsylvania State University, University Park, PA

J. Karl Johnson, Dept.of Chemical and Petroleum Engineering, University of Pittsburgh, Pittsburgh, PA

The reactive force field code ReaxFF differs from traditional force fields in that the bonds are not defined explicitly, but rather defined as a function of manybody interatomic distance, which allows for changes in bond order. This provides a valuable means by which large-scale reactive molecular dynamics can be performed without resorting to expensive quantum chemical ab initio methods. Periodic plane-wave density functional theory (DFT) calculations have been used to derive a ReaxFF parameter set for cobalt. These parameters have been used in a battery of calculations to evaluate the accuracy and performance with respect to other computational methods. We compare ReaxFF with additional DFT calculations, semi-emprical embedded-atom potentials, and experiments, for various large-scale bulk phase defect systems. We compare energetic, structural, and dynamic properties, were possible. In addition to defect systems, ReaxFF is used to examine simple surface behavior such as reconstruction and oxidation, ultimately providing insight into the role of cobalt in catalytic applications.

See more of this Session: First-Principles Simulation of Condensed Phases: Surfaces
See more of this Group/Topical: Computational Molecular Science and Engineering Forum

First-Principles Theoretical Analysis of Doping in II-VI Compound Semiconductor Nanocrystals

Tejinder Singh, Department of Chemical Engineering, University of Massachusetts, Amherst, MA

T. J. Mountziaris, Department of Chemical Engineering, University of Massachusetts, Amherst, MA

Dimitrios Maroudas, Department of Chemical Engineering, University of Massachusetts, Amherst, MA

Nanocrystalline structures of compound semiconductors, such as the II-VI compounds ZnS, CdSe, and ZnSe, exhibit size-dependent optoelectronic properties and form the basis for a new generation of nanoelectronic and photovoltaic devices, as well as biological labels. Doping in semiconductor nanocrystals allows for precise control of their optical and electronic properties. However, in spite of its feasibility, doping of semiconductor nanocrystals has been an extremely difficult task. Self-purification of nanocrystals makes the incorporation of dopants into nanocrystals even more difficult. Typically, Mn-doped ZnSe nanocrystals have been grown using hot-injection organometallic synthesis. It has been observed that increase in anion:cation ratios leads to higher doping efficiencies. This has been attributed to adsorption of Mn dopant atoms into ZnSe nanocrystals through their surface facets, in particular ZnSe(001)-(2x1) surface facets. In this context, we aim at obtaining a fundamental and quantitative understanding of dopant adsorption and diffusion on II-VI semiconductor nanocrystal surfaces, which can help elucidate the mechanisms of dopant incorporation into growing nanocrystals.

In this presentation, we report results on dopant adsorption and diffusion on surface facets of ZnSe and $ZnSe_{1-x}S_x$ nanocrystals based on first-principles density functional theory (DFT) calculations within the generalized gradient approximation. In our DFT calculations, we have employed slab supercells, plane-wave basis sets, and the projector-augmented wave method. We have also implemented the climbing-image nudged elastic band method to construct fully optimized dopant diffusion pathways. We have made indirect comparisons between our DFT analysis and experimental findings on doping efficiencies.

We have computed the surface energies of the three high-symmetry surfaces [(001), (110), and (111)] of ZnSe and examined the corresponding surface reconstructions and their stability as a function of the anion (Se) chemical potential. Based on these findings, we have constructed the equilibrium crystal shape (ECS) of ZnSe nanocrystals as a function of the Se chemical potential and found that anion-rich reconstructed surface facets are present in the ECS of a ZnSe nanocrystal. We have computed the binding energies of Mn dopant atoms onto all of the ZnSe nanocrystal facets and found that they depend strongly on the surface morphology and nanocrystal shape. We conclude that all anion-rich surfaces contribute toward dopant adsorption onto ZnSe nanocrystal surface facets.

In addition, our DFT calculations indicate that the binding energy for Mn adsorption onto various sites of all surface facets increases with increasing dopant surface concentration. This low binding energy at low dopant surface concentration explains the doping difficulties during nanocrystal growth. Furthermore, we have analyzed several dopant migration pathways for Mn diffusion on the ZnSe(001)-(2x1) surface. The calculated activation barriers for migration along the Se dimer rows range from 0.17 eV to 0.43 eV, depending upon the dopant surface concentration. Due to the low activation barriers, dopant atoms can migrate fast along the Se dimer rows without substantial surface relaxation. However, migration of the Mn atom across the Se dimer rows from the dimer site (Mn adsorbed onto the Se dimer) to the trough site (Mn adsorbed in the trough between Se dimers) is governed by a high-barrier pathway. Finally, the above analysis of dopant adsorption and diffusion on nanocrystal facets has been extended to $ZnSe_{1-x}S_x$ nanocrystals in order to examine the effects of nanocrystal alloying on the doping efficiency.

See more of this Session: First-Principles Simulation of Condensed Phases: Surfaces
See more of this Group/Topical: Computational Molecular Science and Engineering Forum

The Configurational Correlation: Relating Ab Initio Energies of Surface Oxides Across Transition Metal Surfaces

Spencer D. Miller, Chemical Engineering, Carnegie Mellon University, Pittsburgh, PA

John R. Kitchin, Department of Chemical Engineering, Carnegie Mellon University, Pittsburgh, PA

The correlation between DFT based adsorption energies of single atomic adsorbates across different transition metal surfaces has previously been demonstrated. The source of this near linear correlation is a shared electronic structure effect coverage dependence that can be observed through the modification of the surface metal's d-band structure caused by the presence of adsorbates.

One such single atomic adsorbate that follows this configurational correlation on transition metal surfaces (such as Ag, Au, Pt, and Pd) is oxygen. At low oxygen exposures chemisorbed oxygen can be observed on these surfaces, but at higher exposures, but not high enough to form bulk oxides, surface oxides are either observed or predicted. These surface oxide configurations consist of metal ad atoms sitting atop the surface in addition to adsorbed oxygen. These configurations can play an influential role in the phase behavior of these systems under a range of oxygen partial pressures, and a weakness of some ab initio studies into oxygen's phase behavior on these surfaces has been neglecting these configurations.

We show that the configurational correlation can be extended to these surface oxide configurations by examining the energies of several simple surface oxide configurations as well as several additionally experimentally relevant configurations and showing that these energies correlate across different transition metal surfaces. We also investigate the phase behavior of these surface oxide configurations and how their stability relates to the chemisorbed oxygen configurations. Lastly, we also investigate whether a common bonding mechanism can be found as the source of this correlation.

See more of this Session: First-Principles Simulation of Condensed Phases: Surfaces
See more of this Group/Topical: Computational Molecular Science and Engineering Forum

Paper 517a

Utilization of Molecular Simulations in Aerospace Materials: Simulation of Thermoset Resin/Graphite Interactions

Andrea R. Browning, The Boeing Company, Seattle, WA

Introduction

Aerospace structural materials include a variety of high performance materials from metal alloys to composites. The performance capabilities of structural materials and airplane design are necessarily linked. While the continued development of new materials has allowed for new airplane designs, designs are limited to available materials. It is desirable for the development of new materials to be driven by the requirements of new designs, or "materials by design". A "materials by design" process requires an understanding of structure/property relationships. Computational materials modeling, including molecular simulations, is one tool available to elucidate structure/property relationships and enable "materials by design". For this reason, Boeing has integrated molecular simulations into the materials design process.

An important class of aerospace structural materials is carbon fiber/thermoset resin composites. The performance capabilities of the thermoset resin are critical to the overall capabilities of the composite material. In the design of new thermoset resins, both a quick screening method to determine new resins that may meet performance criteria and a method that increases fundamental understanding of structure/property relationships is desired. Molecular simulations of thermoset resins can be used in both capacities.

Molecular simulations must use accurate descriptions of the interactions between individual components (forcefields) and representative depictions of the material's chemical structure (molecular model). Examples of molecular simulation of thermoset resin at Boeing and in academia can be found in the literature. The method for thermoset resin modeling developed at Boeing[1] will be described in the **Methods** section. Early modeling of cross-linking polymers utilized coarse-grained Monte Carlo simulations of network formation[2,3] and did not always include topological effects[4,5]. The objective of these models was to calculate structural properties such as molecular weight distribution and gel fraction. While adding to the understanding of cross-linking processes, these models could not provide detailed information about the relationship between monomer structure and resulting properties.

Later models of cross-linked polymers included full atomistic detail and utilized molecular dynamics simulations. The cross-linked structures were formed by a variety of methods including joining reaction groups when they are within a prescribed distance[6-8], sequentially reacting groups with the closest proximity[9], and cross-linking in a coarse-grained Monte Carlo simulation followed by reverse mapping to full atomistic detail[10]. Simulations (molecular dynamics and/or minimization) were also included between reaction steps in some of the methods. For a summary of the methods see Ref [8].

Mechanical and thermal properties have been predicted using atomistic models of thermoset resins. For example, Fan and Yuen[11] found good agreement between simulation and experimental glass transition temperature and Young's modulus for EPON 862 epoxy with triethylenetetramine cross-linker. Fan and Yuen did not include details about their cross-linking procedure but did use molecular dynamics simulations for the property calculations. Lin and Khare[12] also used molecular dynamics simulations to calculate thermal properties of diglycidyl ether bisphenol A epoxy with trimethylene glycol di-*p*-aminobenzoate cross-linker both with polyhedral oligomeric silsequionane (POSS) molecules included in the network and without POSS. The resulting simulation glass transition temperature for the resin without POSS was higher than experiments by 6%. This was expected given the relatively high cooling/heating rate used in the simulations. The inclusion of POSS in the network illustrates that nanocomposites containing nanofillers incorporated into thermoset resins can also be modeled using atomistic techniques.

In addition to calculation of mechanical and thermal properties, estimation of interaction properties between polymers and various surfaces is also desired. For example Yarovsky and Evans simulated epoxy resins on alumina surfaces and found that the number of free hydroxyl groups per unit surface area affects the strength of the surface adhesion[7]. Henry and coworkers simulated polyester interacting with graphite and silica surfaces and found that the adhesion between polyester and graphite is stronger than polyester and silica[13]. Interaction properties between thermoset resins and graphite are of particular interest for aerospace materials given the affect interfacial adhesion between carbon fibers and polymer matrix has on composite properties[14,15].

This presentation will discuss the simulation method used by Boeing to model thermoset resins. Both the preparation of the thermoset resin "molecule" and simulation cell will be included. The interaction of a model thermoset resin with graphite will also be explored. Results of interaction energy calculations and uni-axis tension test will be given. The presentation will conclude with a discussion of resin/graphite interaction results and summary of thermoset modeling at Boeing.

Methods

A model that accurately captures the salient components of thermoset resins but is computationally tractable is necessary for the use of computational materials methods in thermoset resin development. In general, the thermoset resins of interest are epoxy resins with amine curing agents. A computational model of thermoset resins with atomistic details (i.e. atoms are included in the model but not electrons) is desired so that the affect of monomer chemical structure can be evaluated. As described by Christensen in Ref [1], thermoset resins are modeled as dendrimers, or highly branched polymers, based on the theory of Flory[16] and Stockmayer[17]. The dendrimers do not contain intramolecular cyclic structures since the rigidity and size of the epoxy monomers make it unlikely for rings to form[18].

Reactions are not directly modeled in the simulations, so the monomer structure must be modified to approximate the after reaction state, see **Figure 1**. The epoxide ring on the epoxy monomer is opened, and a hydrogen atom is added to the epoxide oxygen creating an alcohol group. In the modified monomers, hydrogen atoms are used as place holders where the connection between epoxy monomer and curing monomer will be made. Once a connection between amine and epoxy monomers has been made, the place holder hydrogen atoms are removed. The monomers are further amended so that charge groups (collections of connected atoms with net charge of zero) in the fully connected dendrimer are not split between monomers. This allows for charge group cut-offs to be used in the non-bonded energy calculations[19]. Amine secondary reactions can be accounted for by making similar modifications as those shown in **Figure 1** to the second amine reaction site.

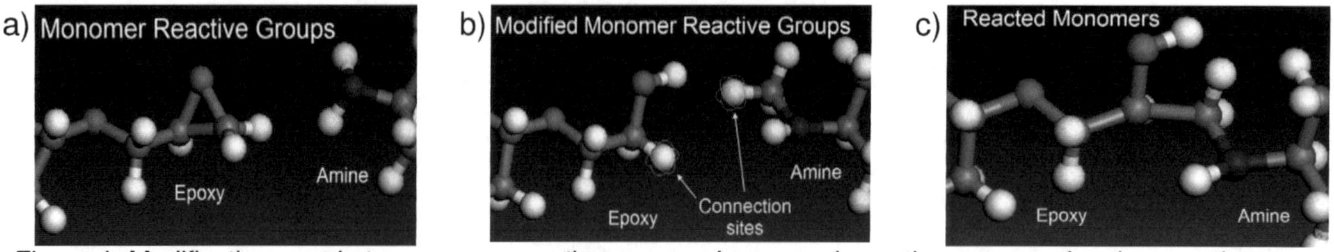

Figure 1: Modifications made to monomer reactive groups. Image a shows the epoxy and amine reactive groups without modification. Image b shows the reactive groups modified with the epoxide ring opened and a hydrogen atom added to the epoxide oxygen to approximate the after reaction state, a carbon moved from the epoxy to the amine so that charge groups in the connected dendrimer are not split between monomers, and hydrogen atoms added as place holders where the connection between monomers will be made. Image c shows the reacted (connected) monomers.

The dendrimers are constructed in Materials Studio by Accelrys Software Inc. using the Dendrimer Building Tool[20]. The building tool constructs the dendrimers by connecting epoxy monomers to an amine monomer seed, then adding successive generations of amines and epoxies, see **Figure 2**. The stoichiometry of the dendrimer can be controlled by disabling some of the available connection sites during the addition of selected generations ("defect creation"). Dendrimers modeled generally contain approximately 3000 - 6000 atoms

Figure 2: Five generation (including amine seed) dendrimer. Amine monomers are shown in gray and epoxy monomers in red. The epoxy is two functional and the amine is four functional. The structure does not contain defects.

After the dendrimer is constructed, the Amorphous Cell module[20] by Materials Studio is used to place the dendrimer in a simulation cell with periodic boundary conditions. The COMPASS force field[20] with group based cut-offs for non-bonded energy calculations are used in the cell construction and all subsequent simulations unless otherwise noted. The true structure of thermoset resins at an atomistic level is inhomogeneous given that spatial fluctuations in concentration exist in the pre-cured mixture of monomers. This inhomogeneity can be partially captured by creating a statistical set of different simulation cells (i.e. different cells constructed by the Amorphous Cell module), performing the equilibration simulations and property calculations described below for each simulation cell, and averaging the resulting properties.

The simulation cells are created at a low density and high temperature (650 K). The simulation cell must then be annealed and compacted by a series of NVT (constant number of atoms, volume, and temperature) and NPT (constant number of atoms, pressure, and temperature) molecular dynamics simulations with the Discover module of Materials Studio. Finally, the simulation cell is minimized, and a 0.25 ns NPT molecular dynamics simulation at 300 K and 0.0001 GPa (room temperature and pressure) is performed. **Figure 3** shows a simulation cell after the 0.25 ns simulation. The 0.25 ns simulation equilibrates the structure at the desired conditions and is used to estimate the bulk density of the thermoset. The simulated density of diglycidyl ether bisphenol A epoxy with a polypropylene oxide triamine cross-linker at various stoichiometries was found to agree well with experiments[1]. Other bulk properties can be estimated by running subsequent simulations. For example, the elastic constants can be estimated by performing a static elastic constants calculation[21] on a selected number of frames (simulation cell configurations) from the end of the 0.25 ns simulation.

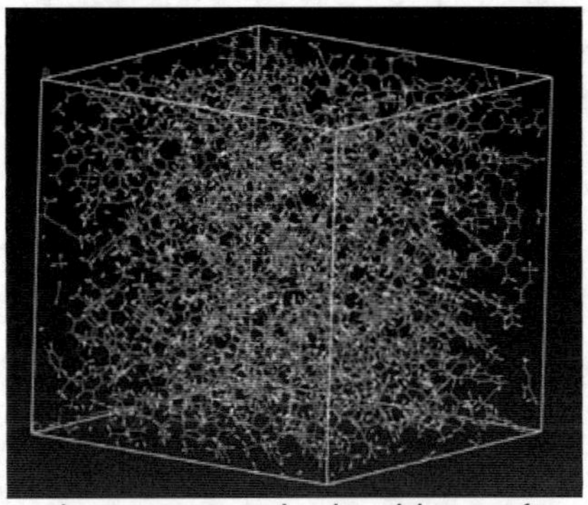

Figure 3: Simulation cell containing a thermoset resin dendrimer after 0.25 ns NPT molecular dynamics simulation.

The simulation cell after the 0.25 ns simulation can also be used to explore interactions with different substrates. In this presentation, interactions with graphite will be described. The simulation cell used to stimulate interactions with graphite (hereafter called the layered cell) is constructed by stacking in the z direction the thermoset resin simulation cell with a cell containing graphite in the desired orientation. The periodic boundary conditions in the z direction are removed from the thermoset resin simulation cell before stacking. The z-dimension of the layered cell is adjusted so that it is the sum of the resin simulation cell z-dimension and graphite cell z-dimension, thereby keeping the density of the graphite and resin constant. After construction the layered cell is equilibrated by running a 0.1 ns or longer NVT simulation. Equilibration is determined by monitoring the potential energy of the layered cell. **Figure 4** shows a schematic of the layered cell.

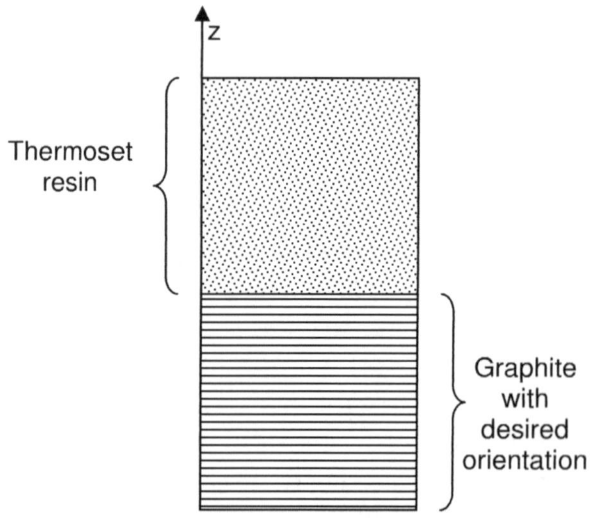

Figure 4: Schematic of the layered cell.

Two different simulations are preformed after equilibration to evaluate the adhesion between the thermoset resin and graphite. The first is a uni-axis tension simulation where the tension is applied in a constant strain method. In this simulation, the cell is extended in the z direction; thereby, applying tension to the perpendicular to the resin graphite interface, see **Figure 5a**. The simulation is composed of a series of small extensions (less than 2%) with 0.01 ns NVT molecular dynamics simulations after each extension. As tension is applied, voids most form in the resin, graphite, or at the interface. Where the voids form indicates where the lowest adhesion forces are found.

The second simulation is an interaction energy calculation. In this calculation a large vacuum space is added to the layered cell so that the interaction energy of only one interface is calculated. The interaction energy (ΔE) is calculated as the difference between the total energy of the layered cell (E_{total}) and the energy of the resin and graphite alone ($E_{resin}, E_{graphite}$)[22], see **Equation 1**. Interaction energy is generally reported as energy per area, so ΔE is divided by the area of the interface[13]. The interaction energy is calculated every 0.0025-0.001 ns for the last 0.025 ns of a 0.1 ns NVT molecular dynamics simulation at 300K. For the interaction energy calculation, Ewald summation method[19] is used for non-bonded energy calculations rather than group based cut-offs.

$$\Delta E = E_{total} - \left(E_{resin} + E_{graphite}\right) \qquad \text{Equation 1}$$

Results

A model resin containing a diglycidyl ether bisphenol epoxy with a diamino diphenylsulphone (DDS) cross-linker was used to examine the resin/graphite interactions. Graphite with ABAB stacking (hexagonal graphite) was used exclusively. The graphite can be aligned relative to the resin/graphite interface in many different manners. In this study, the graphite basal sheets were aligned so they were parallel to the graphite/resin interface (hereafter termed a [0 0 1] graphite surface). Layered cells with other graphite orientations were also simulated but the results will not be discussed here. A uni-axis tension simulation as described in the **Methods** section with the tension applied perpendicular to the graphite/resin interface (see **Figure 5a**) was performed. After a total strain of approximately 30%, the resin remained closely associated with the graphite surface and the average graphite basal plane spacing remained constant (less than 1% change). However voids were observed in the resin, see **Figure 5b**. The same results were found if the Ewald summation method for non-bonded energy calculation was used. The results were also consistent when a layered cell containing approximately twice as many atoms was simulated. The formation of voids in the resin indicates that cohesive failure in the resin occurs before adhesive failure. Cohesive failure is the desired mode of failure[23]; therefore, no changes need to be made to the thermoset resin with [0 0 1] graphite surface system to encourage the desired mode of failure.

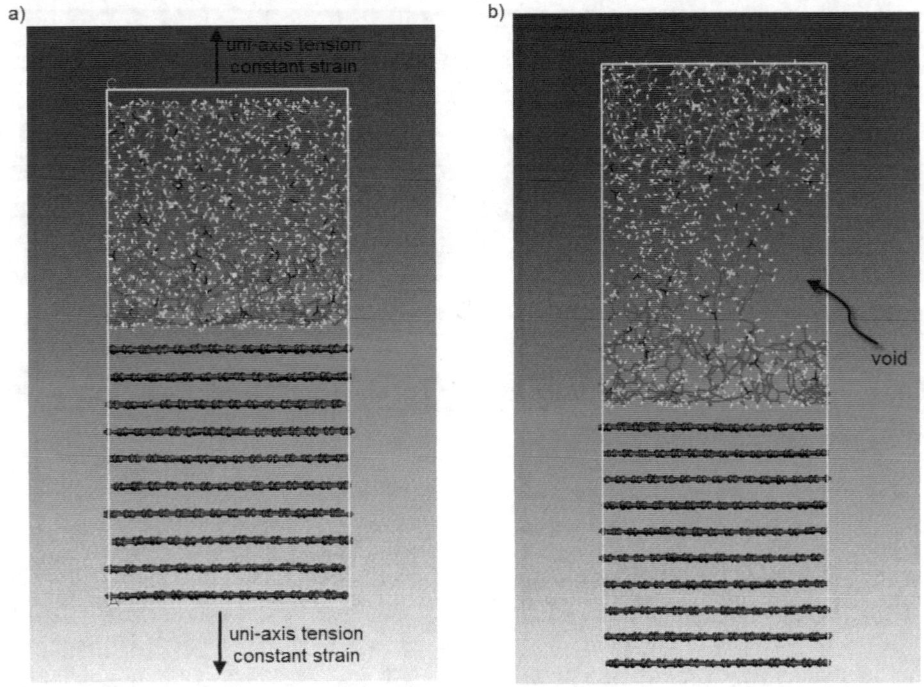

Figure 5: Layered cell with thermoset resin and [0 0 1] graphite surface before uni-axis tension (part a) and after ~30% strain (part b).

An interaction energy calculation was also performed for the model resin with [0 0 1] graphite surface. For comparison, the interaction energy between two thermoset resin dendrimers was calculated along with the interaction energy between graphite basal sheets. The results of all the interaction energy calculations are shown in **Table 1.** The magnitude of the interaction energy between the thermoset resin with [0 0 1] graphite surface was higher than the thermoset resin with itself. This supports the results of the uni-axis tension simulation.

Table 1: Summary of the interaction energy calculations.

	Resin/ [0 0 1] Graphite	Resin/Resin	[0 0 1] Graphite/ [0 0 1]Graphite (interaction between graphite basal sheets)
$\Delta E / A$ (kcal/molÅ^2)	-0.33	-0.23	-0.80

Summary

Computational materials modeling has allowed Boeing to begin to build a "materials by design" process. An atomistic level method to model thermoset resins has been develop that allows for the quick screening of new resins and the exploration of structure/property relationships. The method models thermoset resins as dendrimers, rather than constructing a 3 dimensional network from a simulation cell of monomers. Though this model does not rigorously capture the structural variation present in a true thermoset resin, it has been shown to be sufficient to approximate the bulk properties of thermoset resins.

The thermoset resin modeling methods can also be extended to probed thermoset resin properties such as interaction with substrates. As an example, the interaction of a thermoset resin with graphite were simulated and discussed. It was found that the interaction between the thermoset resin and [0 0 1] graphite surface is stronger than the interactions of the thermoset resin with itself, and thermoset resin cohesive failure occurs before resin/graphite adhesive failure. This implies that in the case of the model thermoset resin with [0 0 1] graphite surface, no changes need to be made to the system to encourage cohesive failure over adhesive failure. This recent work in material simulations demonstrates that the future of aerospace materials development includes simulation tools.

Acknowledgements

The author would like to thank Steve Christensen for instructions on the thermoset resin modeling method and helpful discussions.

References

1. Christensen, S. (2007), "Atomistically Explicit Molecular Dynamics Simulations of Thermosetting Polymers," *in Proc. of 39th ISTC SAMPE Conf.,* Cincinnati, OH.
2. Leung, Y.-K. and B.E. Eichinger (1984), "Computer Simulation of End-Linked Elastomers. I. Trifunctional Networks Cured in the Bulk," *J. Chem. Phys.*, **80**, 3877-3884.
3. Shy, L.Y., Y.K. Leung, and B.E. Eichinger (1985), "Critical Exponents for Off-Lattice Gelation of Polymer Chains," *Macromolecules*, **18**, 983-986.
4. Cheng, K.-C. and W.-Y. Chiu (1994), "Monte Carlo Simulation of Polymer Network Formation with Complex Chemical Reaction Mechanism: Kinetic Approach on Curing of Epoxides with Amines," *Macromolecules*, **27**, 3406-3414.
5. Chen, Y.-C. and W.-Y. Chiu (2000), "Polymer Chain Buildup and Network Formation of Imidazole-Cured Epoxy/Phenol Resins," *Macromolecules*, **33**, 6672-6684.
6. Doherty, D.C., B.N. Holmes, P. Leung, and R.B. Ross (1998), "Polymerization Molecular Dynamics Simulations. I. Cross-Linked Atomistic Models for Poly(methacrylate) Networks," *Comput. Theor. Polym. S.*, **8**, 169-178.
7. Yarovsky, I. and E. Evans (2002), "Computer Simulation of Structure and Properties of Crosslinked Polymers: Application to Epoxy Resins," *Polymer*, **43**, 963-969.

8. Varshney, V., S.S. Patnaik, A.K. Roy, and B.L. Farmer (2008), "A Molecular Dynamics Study of Epoxy-Based Networks: Cross-Linking Procedure and Prediction of Molecular and Material Properties," *Macromolecules,* **41**, 6837-6842.
9. Wu, C. and W. Xu (2007), "Atomistic Simulation Study of Absorbed Water Influence on Structure and Properties of Crosslinked Epoxy Resin," *Polymer*, **48**, 5440-5448.
10. Komarov, P.V. et al. (2007), "Highly Cross-Linker Epoxy Resins: An Atomistic Molecular Dynamics Simulation Combined with a Mapping/Reverse Mapping Procedure," *Macromolecules*, **40**, 8104-8113.
11. Fan, H.B. and M.M.F Yuen (2007), "Material Properties of the Cross-Linked Epoxy Resin Compound Predicted by Molecular Dynamics Simulation," *Polymer*, **48**, 2174-2178.
12. Lin, P.H. and R. Khare (2009), "Molecular Simulation of Cross-Linked Epoxy and Epoxy-POSS Nanocomposite," *Macromolecules*, **42**, 4319-4327.
13. Henry, D.J., C.A. Lukey, E. Evans, and I. Yarovsky (2005), "Theoretical Study of Adhesion between Graphite, Polyester and Silica Surfaces," *Mol. Simulat.*, **31**, 449-455.
14. Drzal, L.T. and M. Madhukar (1993), "Fibre-Matrix Adhesion and its Relationship to Composite Mechanical Properties," *J. Mater. Sci.*, **28**, 569-610.
15. Tang, L.-G. and J. L. Kardos (1997), "A Review of Methods for Improving the Interfacial Adhesion between Carbon Fiber and Polymer Matrix," *Polym Composite*, **18**, 100-113.
16. Corezzi, S., D. Fioretto, D. Puglia, and J.M. Kenny (2003), "Light Scattering Study of Vitrification during the Polymerization of Model Epoxy Resins," *Macromolecules*, **36**, 5271-5278.
17. Flory, P.J. (1941), "Molecular Size Distribution in Three-Dimensional Polymers: I, Gelation," *J. Am. Chem. Soc.*, **63**, 3083-3090.
18. Stockmayer, W.H. (1943), "Theory of Molecular Size Distribution and Gel Formation in Branched-Chain Polymers," *J. Chem. Phys.*, **11**.
19. Leach, A.R. (1996), *Molecular Modelling Principles and Applications*, Longman, Harlow Essex, England.
20. Accelrys Software Inc. (2008), *Materials Studio Release Notes, Release 4.4*, Accelrys Software Inc., San Diego, CA.
21. Theodorou, D.N. and U.W. Suter (1986), "Atomistic Modeling of Mechanical Properties of Polymeric Glasses," *Macromolecules*, **19**, 139-154.
22. Gou, J. et al. (2004), "Computational and Experimental Study of Interfacial Bonding of Single-Walled Nanotube Reinforced Composites," *Comp. Mater. Sci.*, **31**, 225-236.
23. Gersappe, D. and M.O. Robbins (1999). "Where do Polymer Adhesives Fail?," *Europhys. Lett.*, **48**, 150-155.

Biomolecular Transport and Adsorption in Porous Polymer Adsorbent Media

E. Riccardi, J.-C. Wang, and A. I. Liapis, Missouri University of Science and Technology, Rolla, MO 65409-1230

The important advances that have taken place in the fields of molecular biology and separation science in the past twenty five years have contributed significantly in the development of large-scale production of diagnostics and biotherapeutics. In downstream purification and separation of bioactive molecules, the capture step serves the dual purpose of removing the bulk impurities and concentrating the product of interest. Because of this, ion-exchange chromatography (IEC) has become a very important separation method in downstream purification for the isolation of peptides, enzymes, and proteins. The IEC affinity groups/ligands are either positively or negatively charged groups and are most often linked to the base matrix via a polymeric extender because of the positive effect caused by the extender and which is extensively discussed in the literature. The porous polymer layer grafted on the surface of the pores of the base matrix increases the effective interactive volume as well as the steric availability of the affinity groups/ligands for the biomolecule to be adsorbed; this in turn could increase the mass transfer rate as well as the total capacity available.

In the process used for the construction of porous polymer adsorbent media, stage I concerns itself with the construction of the porous polymer layers formed from the mutual interaction of the polymeric extender molecules which are grafted on the surface of the pores of the base matrix, while stage II involves the immobilization of the affinity groups/ligands on the surface of the pores of the immobilized porous polymer layers. The adsorption process which involves (i) the transport mechanisms of the charged adsorbate biomolecules of interest in the porous polymeric adsorbent medium and (ii) the interaction mechanisms involved in the adsorption of the adsorbate molecules onto the immobilized affinity groups/ligands, is considered to represent stage III in the study of the separation of biomolecules by ion-exchange adsorption employing polymeric adsorbent particles whose formation involves stages I and II. Recently, Riccardi et al. have presented a methodology which enables one to (i) construct through molecular dynamics (MD) modeling and simulations different porous polymeric media and (ii) characterize the pore structure of the constructed porous polymeric media with respect to distributions of pore size, pore surface areas, pore volume, and pore connectivity. Depending on the size of the desired adsorbate molecule and the size of the desired affinity group/ligand to be immobilized on the surface of the pores of the polymeric structures in order to form an adsorbent medium, the porous polymeric medium having the structural characteristics capable of potentially producing the most effective porous adsorbent medium can be constructed and selected. This work of Riccardi et al. provides a methodology for modeling, designing, constructing, characterizing, and selecting the appropriate porous polymeric structure to be used for the immobilization of the desired affinity groups/ligands and, therefore, it represents a theoretical method for studying and predicting the behavior of stage I.

After the selection of the most promising porous polymeric structure, the surface of the pores of the polymeric medium is then activated and followed by the immobilization of the selected affinity groups/ligands onto the activated pore surfaces commences. In this respect, Riccardi et al. have also recently presented a theoretical approach which enables one to (i) model and study the activation of the pore surfaces and the immobilization of the selected affinity groups/ligands onto the activated pore surfaces through the use of MD modeling and simulation methods, (ii) determine the spatial distributions of (a) the density of the immobilized affinity groups/ligands, (b) the counterions, (c) the local nonelectroneutrality, and (iii) analyze and characterize the structure of the porous adsorbent medium composed of the polymeric chains and the immobilized charged ligands. Furthermore, Riccardi et al. used their methodology to investigate and study the immobilization of charged ligands onto the activated surfaces of the pores formed by the interaction of dextran polymer chains, the spatial distributions of the dextran monomers, charged ligands, and counterions under different ligand loadings, as well as the modifications in the porous dextran polymer structure due to the presence of the charged ligands. Thus, this work of Riccardi et al. represents a theoretical method for studying and predicting the behavior of stage II. The works by Riccardi et al. together have established a methodology able to construct and analyze the structure of different porous adsorbent media and also provided a procedure and criteria for selecting the structure of the porous adsorbent medium that could be particularly effective for the transport and adsorption of certain types of adsorbate biomolecules depending on the size of their hydrodynamic diameter.

Both microscopic and macroscopic modeling studies indicate that the transport and adsorption of adsorbate molecules are expected to be more efficient when the porous adsorbent medium has a high surface area, the size of the pores is sufficient to accommodate properly adsorbate molecules and also facilitate effectively the transport of the adsorbate molecules even after adsorption on the pore surfaces has occurred, and a high pore connectivity so that the adsorbate molecules can be effectively transported from the outer surface of the porous adsorbent medium to the interior porous regions of the medium, and, thus , utilize effectively the immobilized affinity groups/ligands. It is important to mention here that the microscopic modeling studies of Riccardi et al. have shown that the spatial distribution of the density of immobilized charged ligands in ion-exchange adsorbent particles is most likely non-uniform and the adsorbent particles also exhibit local nonelectroneutrality. The concept of non-uniformity in the spatial distribution of the density of immobilized charged ligands was used in a multiscale modeling study where non-uniform functional forms describing the spatial distribution of immobilized ligands in porous spherical adsorbent particles were constructed based on results obtained from microscopic modeling and were then employed in a macroscopic model that could describe the transport and adsorption of a single protein in different porous adsorbent media. The results obtained from the macroscopic model indicate that inner radial humps in the concentration profile of the single adsorbed protein in the particles could occur when the functional forms for the non-uniform spatial density distributions of the immobilized ligands have certain mathematical properties. The results are very interesting and important because inner radial humps in the concentration profile of different adsorbed proteins have been measured experimentally and can have implications with respect to the rates and system efficiencies at which adsorption separations could be performed. Furthermore, the results of the multiscale modeling study have indicated an experimental method by which the spatial density distribution of the immobilized ligands in porous adsorbent particles could be determined, and such measurements could also be used for quality control of the adsorbent medium.

In this work, the transport and adsorption of a charged biomolecule onto the charged ligands immobilized on a representative porous polymeric dextran layer are studied by conducting MD simulations with potential models containing explicit atomistic features. This work is considered to provide a methodology for studying the behavior of stage III. Desmopressin has been selected to be the charged adsorbate biomolecule which is a synthetic analogue of the neurohypophyseal peptide hormone vasopressin and has high antidiuretic and antibleeding potencies. The desmopressin molecule has nine amino acid residues bonded together in the following sequence

S———————————S
| |
Mpr – Tyr – Phe – Gln – Asn – Cys – Pro – D-Arg – Gly – NH_2

and has a hydrodynamic diameter of about 12 Å. Considering the size of desmopressin and other relevant factors, the porous adsorbent medium obtained in our previous work that has an intermediate dextran surface density provides a suitable porous structure for the transport and adsorption of desmopressin because it has high pore connectivity and large in diameter pores (up to 100 Å) which have continuous long lengths along the direction of net transport (z axis) and could possibly accommodate several desmopressin molecules.

The physical characteristics of and the simulation models for the selected porous adsorbent medium having an intermediate dextran density have been presented and explained in detail in our prior work and, thus, only a brief summary is given here. The system is constructed with twenty dextran chains grafted on a non-flat surface that is represented by the following model

$$U_{is}(x,y,z) = 2\pi\varepsilon_{is}\left(\frac{\sigma_{is}}{R_{0s}}\right)^2\left[\frac{2}{5}\left(\frac{\sigma_{is}}{z_s}\right)^{10} - \left(\frac{\sigma_{is}}{z_s}\right)^4 - \frac{\sqrt{2}}{3\left(\frac{R_{0s}}{\sigma_{is}}\right)\left(\frac{z_s}{\sigma_{is}} + \frac{0.61}{\sqrt{2}}\frac{R_{0s}}{\sigma_{is}}\right)^3}\right], \quad (1a)$$

$$z_s = z - a\left[2 + \cos\left(\frac{2\pi}{b}x\right) + \cos\left(\frac{2\pi}{b}y\right)\right], \quad (1b)$$

where U_{is} is the interaction potential energy between the model surface and a particle *i* at (x,y,z) whose size and interaction strength are denoted by σ_i and ε_i. The interaction parameters are determined by the Lorentz-Berthelot mixing rule $\sigma_{is} = \frac{1}{2}(\sigma_i + R_{0s})$ and $\varepsilon_{is} = \sqrt{\varepsilon_i \varepsilon_{0s}}$, where R_{0s} and ε_{0s} represent the equilibrium distance and the interaction strength between agarose monomers. Along the lateral x and y axes of the simulation box, the linear dimensions are 202Å × 202 Å and periodic boundary conditions are applied. Grafted at one end onto the agarose surface, the dextran chains are forty monomers long of which 5% have side branches that are one or two monomers in length. The dextran inter- and intra-chain interactions were treated by a coarse grain M3B model developed by Molinero et al., which represents each glucose monomer with three coarse grain beads that are denoted as B1,B4, and B6 and are located in correspondence to the C1,C4, and C6 carbon atoms. The bending and torsional motions among consecutive M3B beads are regulated by $U_b(\theta) = \frac{1}{2}k_\theta(\theta_i - \theta_0)^2$ and $U_{tor}(\phi) = \sum_{j=1}^{3}\frac{1}{2}B_j\left[1 + \cos(j\phi - \phi_j^0)\right]$, respectively, for proper conformational representation, and the non-bonded interaction between the coarse grain beads is treated by the Morse potential,

$$U_{\mathrm{nb}}\left(r_{ij}\right)=D_{0,ij}\left\{\left[\exp\left(-0.5\alpha_{ij}\left(\frac{r_{ij}}{R_{0,ij}}-1\right)\right)-1\right]^{2}-1\right\} \quad , \tag{2}$$

where r_{ij} is the distance between two dextran beads, $R_{0,ij}$ is the distance of minimum energy $D_{0,ij}$, and α_{ij} is the steepness parameter of the interaction potential. The bond lengths between two bonded beads are constrained by the SHAKE algorithm. The resultant forces and coordinate adjustments are naturally applied to the corresponding C1, C4, and C6 atoms. The values of the parameters of all the potential models discussed here are the same as those presented in our previous studies. While these coarse grain models are employed in current studies for computational efficiency and for structural consistency to enable comparison with previous results, atomistic models embracing van der Waals (vdW) interactions as well as Coulombic charge-charge interactions are needed for the relevant species in order to capture the finer-scale details of the mechanisms responsible for the transport of the charged adsorbate molecules through the electrolytic solution and their adsorption onto the charged ligands in an ion-exchange chromatography medium. A judicious hybrid simulation approach is thus utilized in this work that combines the above coarse grain models with the atomistic models elaborated below by using the modelling of the dextran chains as the convergence point.

In order to properly evaluate the interactions of dextran with atomistically modeled ligand, desmopressin, water, and counterions, the dextran chains need to be simulated with appropriate atomistic features. In this respect, the M3B model provides a method to match and convert between the coarse grain and the full-atomic representations of glucan polymeric chains via a transformation matrix based on idealized equilibrium structure. We have followed this approach in this work and calculated the locations of all the atoms in each dextran chain at every time step through a trans-rotational matrix coupled with the instantaneous coordinates of the B1, B4, and B6 beads. The united atom approach is further employed to fuse –CH3–, –CH2–, and –CH– groups into pseudoatoms with the interaction centers placed at the positions of carbon atoms. The partial charges of the dextran atoms and pseudoatoms were determined according to the study of Lins and Hünenberger who calculated the partial charge distributions for hexopyranose-based carbohydrates under the condition of electrical neutrality for each monomer and entire carbohydrate chain molecules.

Previously, the ligands immobilized on the dextran polymeric chains have been represented by single beads that are bonded to the B4 sites and carry –1e unit charge at their centers of mass. In this work, a molecule with the same charge and similar dimension and mass, namely mercaptopropionic acid (HSCH2CH2COO–), has been selected as the model ligand. The atomistic representations of mercaptopropionic acid and desmopressin are based on the force field developed by Cornell et al. for protein, nucleic acids, and organic molecules using the AMBER package that was also developed by the same group. The H, N, O, S, and C atoms are modeled as individual atomistic interaction sites, while the –CH3–, –CH2–, and –CH– groups are again modeled as united atoms with interaction centers placed at the positions of the carbon atoms. All the atoms carry partial charges and the net charge of desmopressin (adsorbate) is +1e and that of mercaptopropionic acid (charged ligand) is -1e. Each bond in the model desmopressin and ligand molecules is constrained to the equilibrium length by applying the SHAKE algorithm. In addition, in order to retain realistic molecular structures, other intramolecular interactions have also been simulated, including bond angle bending by

$U_b(\theta) = K_\theta(\theta - \theta_{eq})^2$, dihedral torsion by $U_{tor}(\phi) = \sum \frac{V_n}{2}[1 + \cos(n\phi - \gamma)]$, and non-bonded vdW and electrostatic interactions by

$$\sum_j \sum_{k>j+3} \left\{ 4\varepsilon_{jk} \left[\left(\frac{\sigma_{jk}}{r_{jk}} \right)^{12} - \left(\frac{\sigma_{jk}}{r_{jk}} \right)^{6} \right] + \frac{1}{4\pi\varepsilon_0} q_j q_k \left[\frac{1}{r_{jk}} + \frac{(1+B_1)}{2} \frac{r_{jk}^2}{R_c^3} \right] \right\} \quad , \qquad (3)$$

where the second part of Equation (3) also includes a treatment of the long-range electrostatic interactions ($r > r_c = 12$ Å) based on a generalized reaction field method.

In our simulation system, the water molecules not only completely cover the dextran chains but also form a bulk-like aqueous phase above the dextran layer where the transport of desmopressin starts in the simulation system. For this purpose, a total number of 150,000 water molecules are simulated with the TIP4P/2005 model39 which has been shown to be adequate for simulating thermodynamic properties of dense water systems.39,40 This model considers each water molecule to be composed of two hydrogen atoms each carrying +0.5564e, one neutral oxygen atom as a vdW interaction center, and one massless site carrying -1.128e in close proximity to the oxygen atom. The water molecules are simulated as rigid molecules by using the SHAKE28,29 algorithm to constrain the O-H bonds and the distance between the two hydrogen atoms. The counterions surrounding the ligands and solvated in the aqueous phase are considered to be Na+ in this work and their number compensates the charges of the ligands and desmopressin to provide global electroneutrality for the system. The Na+ cations are represented by spherical particles41 with a point charge and a vdW interaction site at the center. It is important to note here that although the electrostatic interactions are much more dominant, all the vdW interactions discussed above are needed in order to prevent artificial overlap due to Coulombic attractions between opposite charges. The vdW parameters used are determined either directly from the literature30-32,35-37 and/or by using the Lorentz-Berthelot mixing rule when necessary.

The MD simulations in this work are carried out using the leapfrog integration algorithm with a time step of 1.0 fs together with a Gaussian thermostat method to control the temperature at 291 K. Because of the hybridization of coarse grain and atomistic models, the selected adsorbent medium needs to be re-equilibrated prior to the start of the adsorption process. In the present work, a progressive re-equilibration strategy involving multiple steps has been used in order to avoid physically implausible strong repulsions and achieve expedition. Water molecules are the first ones to be converted to atomistic model with random initial orientations. During this conversion step, dextran chains, ligands, and counterions are kept frozen and a Monte Carlo based technique is used to adjust water molecular orientations in order to steer the simulation system towards more reasonable states when strong repulsions occur. This re-equilibration step spans 0.5 ns with the integration time step gradually increasing from 0.5 fs to 1.5 fs, and is followed by a step over 0.3 ns during which dextran monomers and ligands are switched to atomistic models but maintained frozen. Afterwards, the adsorbent medium system is further re-equilibrated for another 0.4 ns with the full set of hybrid models discussed above and with every species and degree of freedom unfrozen.

Meanwhile, in a separate simulation cell an atomistically modeled desmopressin molecule is equilibrated over 0.5 ns in a bulk-like phase of 30,000 TIP4P water molecules to acquire conformations similar to those in a finite-bath environment that the biomolecule is subjected to before adsorption. The equilibrated adsorbate biomolecule is then incorporated into the simulated adsorption system at selected locations above the equilibrated porous dextran layer. The initial position of the desmopressin molecule is above the biggest pore presented by the dextran porous structure and the starting z position is 100 Å from the surface of the base matrix which corresponds to the z position of the outermost ligand plus 1.5 times the cut-off distance of the charge-charge interaction. The combination of a dense aqueous phase, porous adsorbent structure, and the affinity ligands surrounded by counterions makes the transport of the desmopressin molecule toward the ligands such a slow process that cannot be easily accessed by usual MD simulations having a large number of simulated particles. To overcome this difficulty, a special measure is taken in this work to expedite the transport of desmopressin as a semi-continuous process through a series of stop points at different z positions. Once the desmopressin molecule reaches a stop point, the system is re-equilibrated by (i) first fixing its center of mass while allowing other species to move freely and (ii) then releasing the control of the xy coordinates of the desmopressin molecule to allow it to move laterally. Afterwards, the desmopressin molecule is free to move 3-dimensionally.

We have observed that the desmopressin molecule is able to approach an oppositely charged ligand but will be repelled back because the ligand is accompanied by counterions that carry the same charge as desmopressin. However, the charge-charge repulsion also perturbs the counterions and then water molecules to change their locations and the resultant new configurations may attract the desmopressin molecule back to be closer to the ligand again. Such a back-and-forth process can repeat several times and eventually is able to displace sufficient counterions to permit adsorption of desmopressin onto the ligand. We also analyze and find that the desmopressin molecule has very significant effects on the porous structure of the dextran polymeric layer and the distributions of the counterions, which however have not been taken into account by existing macroscopic modeling approaches to date.

Atomistic & Multiscale Modeling of PET

Qifei Wang, Department of Chemical and Bimolecular Engineering, University of Tennessee, Knoxville, Knoxville, TN

David J. Keffer, University of Tennessee, Knoxville, TN

Simioan Petrovan, University of Tennessee, Knoxville, TN

Brock Thomas, Eastman Chemical Company

Abstract

Atomistic and coarse-grained simulations of polyethylene terephthalate (PET) oligomers have been performed. Atomistic Molecular Dynamics (MD) simulations of polyethylene terephthalate (PET) oligomers are performed in the isobaric-isothermal (NpT) ensemble at a state point typical of a finishing reactor. The oligomer size ranges from one to ten repeat units. We report thermodynamic properties (density, potential energy, enthalpy, heat capacity, isothermal compressibility, and thermal expansivity), transport properties (self-diffusivity, zero-shear-rate viscosity, thermal conductivity), and structural properties (pair correlation functions, hydrogen bonding network, chain radius of gyration, chain end-to-end distance) as a function of oligomer size. We compare the results with existing molecular-level theories and experimental data. Scaling exponents as a function of degree of polymerization are extracted. The distribution of the end-to-end distance is bimodal for the dimer and gradually shifts to a single peak as the degree of polymerization (DP) increases. The scaling exponents for the average chain radius of gyration and end to end distance are 0.594 and 0.571, respectively, close to the theoretical value for polymer melts of 0.589. The values of the heat capacity, isothermal compressibility, and thermal expansivity agree well with the available experimental data, which is of much longer PET chains. The scaling exponents for the self-diffusivity and zero-shear-rate viscosity are respectively -2.01 and 0.96, close to the theoretical predictions of -2.1 and 1.0.

A coarse-grained potential for PET is developed based on pair correlation functions from the atomistic simulations. A comparison of the structural and dynamic results between the atomistic and coarse-grained simulations is provided.

Acknowledgments

This research was supported by the Eastman Chemical Company. This research project used resources of the National Institute for Computational Sciences (NICS) supported by NSF under agreement number: OCI 07-11134.5.

See more of this Session: Industrial Applications of Computational Chemistry and Molecular Simulation I
See more of this Group/Topical: Computational Molecular Science and Engineering Forum

Transferable Intermolecular Potentials for Carboxylic Acids and Their Phase Behavior

Amir Vahid, Chemical and Biomolecular Engineering, University of Akron, Akron, OH

Neil H. Gray, Chemstations Inc., Houston, TX

J. Richard Elliott, Chemical and Biomolecular Engineering, University of Akron, Akron, OH

Transferable step potentials are characterized for 39 carboxylic acids. The reference potential is treated with discontinuous molecular dynamics, including detailed molecular structure. Thermodynamic perturbation theory is used to interpret the simulation results and to provide an efficient basis for molecular modeling and characterization of the attractive forces. Four steps are used for representation of the attractive forces with only the first and last steps varied independently. The two middle steps are interpolated such that each site type is characterized by three parameters: the diameter, σ, the depth of the inner well, ε1, and the depth of the outer well, ε4. The depths of the attractive wells are optimized to fit experimental vapor pressure and liquid density data. Generally, the vapor pressure is correlated to an overall 43% average absolute deviation (%AAD) and the liquid density to 5%AAD. The deviations tend to be largest for the higher molecular weight acids. These deviations are larger than the errors previously encountered in characterizing organic compounds, but carboxylic acids present exceptional challenges owing to their peculiar dimerization behavior. Simultaneous correlation of vapor pressure, vapor compressibility factor, and phase equilibria of water+carboxylic acids place several constraints on the nature of the potential model, with the parameters of the present model representing a reasonable tradeoff. In the other words, our model represents minimal deviations for vapor pressure, vapor compressibility factor and phase equilibria of all acids simultaneously while varying the parameters σ, ε1, ε4, εCC(dimerizing site bonding energy), εAD(acceptor-donor bonding energy), and KHB(hydrogen bonding volume) for the acid O= and OH site types. The present model is characterized by one acceptor and one dimerizing site on the carbonyl oxygen and one acceptor and one donor site on the hydroxyl oxygen. The acceptor and donor are capable of interacting with water while the dimerizing site is not. With this model, the saturated vapor compressibility factor of acids with seven or fewer carbons is near 0.5 while higher carbon ratios lead to a compressibility factor approaching 1.0.

To compensate for the high vapor pressure deviations of the transferable potential model, a correction is introduced to customize the molecule-molecule self interaction energy. This adaptation results in deviations of 3.1% for vapor pressure of the pure acid database.

In order to validate the behavior of the model for carboxylic acids in mixtures, 33 binary solutions were considered. Acids in this database ranged from formic to hexadecanoic. The average absolute deviation in bubble pressure for aqueous acid systems is 4.4%, 10.5% for acid+acid systems, and 4.7% for acid+n-alkane systems without a customized interaction correction. When applying the correction, deviations were 2.4% for aqueous systems, 2% for acid systems, and 2.8% for acid+n-alkane systems.

See more of this Session: Industrial Applications of Computational Chemistry and Molecular Simulation I
See more of this Group/Topical: Computational Molecular Science and Engineering Forum

Improved Conformational Sampling for Reaction Ensemble Monte Carlo Simulations

Thomas W. Rosch, Chemical and Biolomolecular Engineering, University of Notre Dame, Notre Dame, IN

Wei Shi, National Energy Technology Laboratory, U.S. Department of Energy, Pittsburgh, PA

Jindal K. Shah, Department of Chemical and Biomolecular Engineering, University of Notre Dame, Notre Dame, IN

Edward J. Maginn, Chemical and Biomolecular Engineering, University of Notre Dame, Notre Dame, IN

In many industrial applications it is imperative to understand the reaction behavior of various chemical mixtures. While comprehensive knowledge may be limited by experimental techniques, molecular simulation provides unique complementary insight into reacting systems. Smith et al.1 and Johnson et al.2 independently developed a methodology capable of predicting equilibrium concentrations of reacting species, referred to as reaction ensemble Monte Carlo (RxMC). Yet, as in other Monte Carlo implementations, RxMC can suffer from inefficient sampling for systems of high density and/or high complexity. We introduce the continuous fractional component3 (CFC) method within the RxMC framework which overcomes sampling difficulties by allowing for gradual introduction and subtraction of products and reactants. We apply this method to study the synthesis of gasoline additive methyl-tert-butyl-ether(MTBE) from isobutene and methanol. This presentation will describe how to implement CFC to model simultaneous vapor-liquid and chemical equilibrium as well as demonstrate enhanced efficiency in calculation of equilibrium concentrations.

1 Smith, W.R., Triska, B., J. Chem. Phys. 1994, 100, 3019-3027

2 Johnson, J.K., Panagiotopoulos, A.Z., Gubbins, K.E., Mol. Phys. 1994, 81, 717-733

3 Shi, W., Maginn, E.J., J. Chem. Theory Comput. 2007, 3, 1451-1463

See more of this Session: Industrial Applications of Computational Chemistry and Molecular Simulation I
See more of this Group/Topical: Computational Molecular Science and Engineering Forum

Computational Exploration of Metal Complexes for Stress-Strain Modulation

Berend Christopher Rinderspacher, Army Research Laboratory, Aberdeen, MD

Jan W. Andzelm, Army Research Laboratory, Aberdeen, MD

Robert H. Lambeth, Army Research Laboratory, Aberdeen, MD

Adam M. Rawlett, Army Research Laboratory, Aberdeen, MD

There is a significant interest in designing metallo-supramolecular polymers that have superior elastic properties and can efficiently dissipate energy. Such polymers have organometallic modules with metals coordinated to organic molecules. Optimal polymers with required stress-strain profiles depend, among other factors, on the strength of the organometallic modules. Systematic search for optimal modules remains difficult because of the astronomical number of possible structures. We have formulated an Inverse Molecular Design (IMD) that efficiently predicts optimal architectures of organometallic modules. This IMD method is related to branch-and-bound/tree-search approach for constrained searching of chemical spaces based on an interpolation of property values on a hypercube.

We present calculations on polymer-linked metal complexes for the evaluation of energy dissipation. To this end, we investigate the energy profiles of stretching, bending and shearing of complexes according to external forces exerted on each complex via the attached polymer strands. Zn, Fe and Cr in conjunction with pyridine derivatives were considered in a poly-ethyleneglycol environment with various counterions. We have used a multiscale simulation approach consisting of quantum-mechanical(Semi-empirical and density-functional) methods and molecular dynamics simulations to predict various deformation modes in a realistic polymer environment.

See more of this Session: Industrial Applications of Computational Chemistry and Molecular Simulation I
See more of this Group/Topical: Computational Molecular Science and Engineering Forum

Quantum Chemistry Simulation of the Condensed Phase Reactions of Bisglycinato Copper(II), Overturns Accepted Theory

David A. Gallagher, CAChe Research, Beaverton, OR

Most quantum studies to date on transition metal complexes, including metallo-enzymes, have been limited to gas-phase calculations. However, gas phase calculations do not always reflect observed behavior in solution. Recent algorithm developments now facilitate the accurate prediction of the thermodynamic effects of specific solvents and mixtures and so, have opened up the condensed phase chemistry of organometallics to theoretical study. Quantum chemistry calculations in simulated solvent now offer new insights that contradict previous rationalizations of the properties of bisglycinato copper(II) in aqueous solution. Bisglycinato copper(II) exists as two stable isomers in the solid state at room temperature, cis and trans. When crystals of the cis monohydrate are heated, they first loose water of crystallization then isomerize to the trans isomer. This is in agreement with high-level DFT gas-phase calculations, which indicate that the trans isomer is more thermodynamically stable than the cis, by over 10 kcal/mol. The same high-level calculations also suggest that the gas-phase energy barrier for the cis to trans isomerization is so low that the trans isomer is likely to be formed spontaneously above 100 K. The fact that crystals of the cis isomer are apparently stable at room temperature, has been attributed to the crystal packing forces being sufficiently strong to prevent isomerization occurring in the solid state. Paradoxically, the cis isomer is the product from synthesis in aqueous solution. It is reportedly because the cis isomer is the kinetic product, i.e. cis crystallizes from solution before it can isomerize to the trans isomer. However, as there should be no significant crystal packing forces in solution, this is inconsistent with the calculated low energy barrier that should lead to spontaneous isomerization to trans above 100 K. Calculations in solution phase suggest a different explanation. While DFT gas phase calculations agree that the trans isomer is more thermodynamically stable in the gas-phase, the presence of a high dielectric water solvent field, appears to stabilize the cis isomer more than the trans isomer, to the extent that their relative thermodynamic stabilities (free energies) are reversed in aqueous solution. As a consequence, the author will propose a new hypothesis for the mechanism of cis-bisglycinato copper(II) synthesis.

See more of this Session: Industrial Applications of Computational Chemistry and Molecular Simulation I
See more of this Group/Topical: Computational Molecular Science and Engineering Forum

Calorimetric and Computational Study of the Heat of Reaction for Propylene Oxide Hydrolysis and Ozonolysis

Ayman D. Allian, Global Pharmaceutical Research and Development, Abbott Laboratories, North Chicago, IL

Steve Richter, Global Pharmaceutical Research and Development, Abbott Laboratories, North Chicago, IL

Travis B. Dunn, Global Pharmaceutical Research and Development, Abbott Laboratories, North Chicago, IL

In process safety, the accuracy of the calorimetric measurements is crucial for safe handling and the scale up of the investigated reaction. We demonstrate that measured heats of reaction via calorimetry can be authenticated using computational chemistry. In the current study, experimental heat of reaction were measured for two reactions namely propylene oxide hydrolysis and alkene ozonolysis at 296.15 and 228.15 K respectively. The heat flow profile for ethylene oxide hydrolysis and the ozonolysis of styrene were experimentally measured using the Omnical SuperCRC and Mettler-Toledo RC1 calorimeters. The reaction enthalpies of these reactions were also computed using high-precision CBS-QB3 calculations. For both reactions, there was an excellent agreement between computational and experimental results.

[1] M. A. Matos, S. C. Clara F. M. Victor, J. of Phys. Chem. A (2008), 112(34), 7961-7968.

[2] M. A. Matos, M. S. Margarida P. A. Diana, F. M. Victor, J. F. Liebman, J. of Phys. Chem. A (2008), 112(40), 10053-10058.

[3] M. A. Matos, M. S. Margarida P. A. Diana, F. M. Victor, J. of Phys. Chem. A (2008), 112(34), 7961-7968.

See more of this Session: Industrial Applications of Computational Chemistry and Molecular Simulation I
See more of this Group/Topical: Computational Molecular Science and Engineering Forum

Molecular and Multi-Scale Simulations of CO_2 and N_2 Transport in ZSM-5 Zeolite Membranes with Framework Substitutions

David A. Newsome, Delft University of Technology, DelftChemTech, Delft, Netherlands

Anton J. Dammers, Delft University of Technology, Delft, Netherlands

Marc-Olivier Coppens, Isermann Department of Chemical and Biological Engineering, Rensselaer Polytechnic Institute, Troy, NY

Zeolites are crystalline aluminosilicates with an ordered network of micropores. Their uses include membrane separations, catalysis, and sequestration. The zeolite ZSM-5 contains a regular array of straight channels in the y-direction, and zigzag channels in the x-z plane. The pores have a diameter of ~5.5 Å. The zeolite lattice has a nominal composition of SiO_2 but the Si atoms can be replaced by Al atoms, along with extra framework cations, such as Na^+, to maintain charge balance. One possible application of ZSM-5 is the membrane separation of CO_2 from flue gases. The molecules CO_2 and N_2 have different kinetic diameters and a different affinity for the zeolite surface. Therefore, they adsorb into and diffuse through the zeolite pores at different rates, which allow them to be separated in a membrane operation. We considered the cases of 0 and 1 Al substitution of Si in the ZSM-5 unit cell to study the role of heterogeneity. The type and concentration of the cation that accompanies the Al is a property that can be controlled via ion exchange to tune the adsorptive and diffusive properties of molecules in ZSM-5. Both N_2 and CO_2, but especially CO_2, have a strong quadrupolar interaction with the Na^+ cation. We have used molecular dynamics (MD) and Grand Canonical Monte Carlo (GCMC) simulations of CO_2 and N_2 diffusing in ZSM-5, with some of the Si atoms substituted by Al- and Na^+ counterions. The MD simulations were run for 20-26 ns at $T = 200$, 300 and 400 K. Diffusivities and residence times around preferred sites were calculated from the simulated trajectories. The diffusivity of CO_2 is of the order of 10^{-6} - 10^{-5} cm^2 / s, while the diffusivity of N_2 is of the order of 10^{-5} - 10^{-4} cm^2 / s. The residence times for CO_2 are in the typical range of 30 - 100 ps when no strong sites are present, but dramatically increase to 1 - 2 ns with Na^+ present. The Na^+ causes an increase in the diffusivity at low loading, until crowding effects cause a decrease. The activation barrier for CO_2 strongly increases due to the presence of Na^+, with the diffusivity of CO_2 showing a maximum. GCMC simulations were used to determine the adsorption isotherms, showing a substantial increase of CO_2 loading with Na^+. content. Though the N_2 molecules are more mobile, they adsorb less strongly than CO_2. To cope with the broad range of time and length scales of membrane transport, we use dynamic Monte-Carlo simulations and mean-field theory of diffusion on a site lattice model representing the pore network topology. These coarse-grained simulations make use of the residence times on the sites and the adsorption isotherms calculated using the just described atomistic models. This multi-scale approach allows a calculation of the selectivity of CO_2 over N_2 and the throughput of a membrane. Current results show a strong dependence of the permeances with the level of heterogeneity. Increasing the Na^+ content typically decrease the permeance at a same, desired level of selectivity, since the CO_2 and N_2 molecules diffuse at a slower rate. The trends of the permeances agree well with experimental measurements, though the quantitative predictions are higher in this defect-free membrane model. Experimental permeance measurements are suspect to intercrystalline barriers, which cause a decrease in the observed permeances. The selectivity of the ZSM-5 membranes is always predicted to be in favor of CO_2, again in agreement with prior experimental observations. Heterogeneity plays an important role in activated diffusion. The residence times and diffusivities indicate that Na^+ cations can dramatically affect the behavior of CO_2 and N_2

molecules, allowing for a more rational basis for membrane design and operation.

See more of this Session: Multi-Scale Modelling I
See more of this Group/Topical: Computational Molecular Science and Engineering Forum

Linking Diffusion Map Variables with Equation-Free Multiscale Atomistic Computations

Ioannis G. Kevrekidis, Department of Chemical Engineering, Princeton University, Princeton, NJ

Benjamin Sonday, Pacm, Princeton University, Princeton, NJ

Mikko Haataja, Mechanical Engineering, Princeton University, Princeton, NJ

Developing effective descriptions of the microscopic dynamics of many physical phenomena can both dramatically enhance their computational exploration and lead to a more fundamental understanding of the underlying physics. Obtaining a good set of variables (macroscopic observables) becomes an important step of such a modeling task.

We will illustrate this through the problem of a driven interface in the presence of mobile impurities, for which an effective reduced model was previously developed based on an Ising variant model and a single empirical coarse variable. The model was partially successful, yet it underlined the necessity of selecting additional coarse variables in certain parameter regimes.

We now use a data mining approach to help identify the coarse variables required.

We discuss the implementation of this Diffusion Map approach, the selection of a similarity measure between system snapshots required in the approach, and the correspondence between empirically selected and automatically detected coarse variables.

We conclude by illustrating the use of the diffusion map variables in assisting the atomistic simulations, and we discuss the translation of information between fine and coarse descriptions using lifting and restriction operators

See more of this Session: Multi-Scale Modelling I
See more of this Group/Topical: Computational Molecular Science and Engineering Forum

From Electrons to Reactive Flow: Coupling First-Principles Statistical Mechanics with Continuum Level Simulations

Sebastian Matera, Theory, Fritz-Haber-Institut der Max-Planck-Gesellschaft, 14195 Berlin, Germany

Karsten Reuter, Theory, Fritz-Haber-Institut der Max-Planck-Gesellschaft, 14195 Berlin, Germany

For the simulation of in-situ experiments in heterogeneous catalysis different coupled aspects have to be encountered. On the hand the surface reactivity depends on the partial pressures and the temperature at surface. On the other hand gas flow and the heat transport in the sample is effected by the consumption and production of species at the surface, and the partial pressures and the temperature at the surface will differ from the applied ones. In order to describe the coupling of both effects we combine first-principles kinetic Monte Carlo simulations [1] to model the surface reactivity with continuum level simulation for the gas and solid phase transport. We apply our method to stagnation flow experiments for the CO oxidation at a RuO2(110) single crystal surface. We show how the gas/solid phase transport affects the outcome of the experiment for different reactor setups (e.g. different flow rates). Especially we investigate the influence of the gas and solid phase transport on the observed turn-over frequency and surface population.

[1] K. Reuter and M. Scheffler, Phys. Rev. B 73, 045433 (2006)

See more of this Session: Multi-Scale Modelling I
See more of this Group/Topical: Computational Molecular Science and Engineering Forum

An Accurate Multiscale Simulation Algorithm for Spatially Inhomogeneous Chemical Reaction Systems

Krishna A. Iyengar, Theoretical & Applied Mechanics, Cornell University, Ithaca, NY

Leonard A. Harris, Department of Computational Biology, University of Pittsburgh School of Medicine, Pittsburgh, PA

Paulette Clancy, Cornell University, Ithaca, NY

There is a great need for accurate and efficient computational approaches that can account for the discrete and stochastic nature of chemical interactions in spatially inhomogeneous systems. This is particularly true in biology and nanoscale materials science, where the common assumptions of deterministic dynamics and well-mixed reaction volumes often break down. Towards this end, we present a spatial version of the "partitioned leaping algorithm" (PLA), a multiscale accelerated-stochastic simulation approach that seamlessly incorporates exact stochastic, τ-leaping and continuum methods into a single algorithmic framework. In developing the approach, we pay special attention to the details of the implementation within the context of a discretized grid, particularly as it pertains to the time step calculation procedure. We discuss conceptual errors that have been made in previous implementations of spatial τ-leaping and illustrate the manifestation of these errors through numerical examples. We also study the high computational cost associated with including an exact-stochastic description in the multiscale framework and consider the effects on the accuracy of the method if it is excluded. Finally, we discuss the fundamental difficulties associated with incorporating efficient exact-stochastic techniques, such as the next-subvolume method, into a spatial leaping approach and suggest possible strategies for overcoming them.

See more of this Session: Multi-Scale Modelling I
See more of this Group/Topical: Computational Molecular Science and Engineering Forum

Molecular Modeling of the Gas Separation in Hybrid Inorganic-Organic Polymeric Membrane

Zhenxing Wang, Chemical Engineering, Virginia Polytechnic Institute and State University, Blacksburg, VA

Luke Achenie, Chemical Engineering, Virginia Polytechnic Institute and State University, Blacksburg, VA

S. Ted Oyama, Chemical Engineering, Virginia Polytechnic Institute and State University, Blacksburg, VA

In industry polymeric membranes are widely used for gas separations. To design membranes with high selectivity, excellent thermal stability and good permeability, multiple types of materials have been investigated. However, it is not clear the physics behind bulk gas separation and the active mechanisms at the molecular level. To address the latter, we employ molecular modeling to study gas separation in the membrane; this understanding will help us predict separation related properties of the membrane at the bulk level. Specifically, the hybrid inorganic-organic membrane, which has silicon dioxide as the sieve and organic groups dispersed, is simulated using molecular dynamics (MD).

We proceed as follows. A bulk alpha crystal of silicon dioxide with some of the atoms removed randomly is used as the starting structure. The BKS force field is used within the MD to convert the crystal structure to the amorphous type. Phenyl groups are then systematically attached to the silicon atoms. Next, the Dreiding force field is used to get the equilibrium conformation of the hybrid inorganic-organic membrane. To investigate the performance of the membrane, the separation of CO2 and CH4 mixture is simulated using non-equilibrium molecular dynamics (NEMD). The selectivity, density of two types of gases in the membrane session, the permeability and other properties are then calculated at different pressures and temperatures. The results are compared with our experimental data.

We are currently modifying the organic groups, changing the radius of the hole and processing other kinds of gas mixtures to see if we can reproduce what we have obtained experimentally. The final intent is to use the MD to predict favorable experimental conditions.

See more of this Session: Multi-Scale Modelling I
See more of this Group/Topical: Computational Molecular Science and Engineering Forum

Improved Analysis of Single-Molecule Force-Spectroscopy Data for Extracting Free Energy Barrier Heights and Barrier Crossing Rates

Arijit Maitra, Department of Nanoengineering, University of California at San Diego, La Jolla, CA

Gaurav Arya, Department of Nanoengineering, University of California at San Diego, La Jolla, CA

Applying controlled amount of forces to single molecules like DNA, RNA, and proteins by means of micromanipulation techniques such as optical tweezers and AFM can provide significant information about the underlying energy landscape. These experiments are generally performed at fixed or constantly increasing forces and the dynamic response of the molecule is recorded in the form of molecular extension vs. force and extension vs. time plots. This "nonequilibrium" response may then be analyzed to extract "equilibrium" properties such as the intrinsic free energy barrier (&Delta G^*), barrier distance (x^*) and transition rate (k_0) of the molecule undergoing conformational transition between different macroscopic states. Recently, Dudko et al. (PRL, 2006; PNAS, 2008) have developed a sophisticated approach based on Kramer's theory of barrier crossing to extract the free energy barrier heights and locations as well as the rates of barrier crossing from force-spectroscopy data in which the molecule is stretched at a constant velocity. Here we present an extension of this approach that provides more accurate estimates of free energies and kinetic rates by additionally exploiting the compliance of the device itself. We also present a validation of this approach by studying a simplified system composed of a Brownian particle in a double-well potential. We are now applying this approach to determine the free energy profiles and kinetics of conformational transitions in coarse-grained models of proteins and RNA.

See more of this Session: Multi-Scale Modelling I
See more of this Group/Topical: Computational Molecular Science and Engineering Forum

Single-Event MicroKinetic Modeling of *n*-Hexane hydroisomerization on Pt/H-ZSM-5

I. Roy Choudhury[a], J.W. Thybaut[a,*], J.F.M. Denayer[b], J.A. Martens[c] and G.B. Marin[a]

[a] *Laboratory for Chemical Technology, Ghent University, Krijgslaan 281 – S5,Gent, B-9000*

[b] *Department of Chemical Engineering, Vrije Universiteit Brussel, Pleinlaan 2, Brussels, B-1050*

[c] *Centre for Surface Chemistry and Catalysis, KU Leuven, Kasteelpark Arenberg 23, Leuven, B-3001*

Abstract

A single-event microkinetic model has been developed for n-hexane hydroisomerization on Pt/H-ZSM-5 considering simultaneous effects of physisorption, diffusion and reaction. Shape selectivity is accounted for through the distinctive physisorption of linear and branched alkanes and multi-component diffusion. The diffusion phenomenona at the molecular level are translated to the crystallite scale and coupled with single-event microkinetics. The concentration dependence of the diffusion coefficients in the presence of acid sites is related to catalyst descriptors. The latter allow the quantitative description of the catalyst properties on the concentration profile on the crystallite scale and n-hexane conversion. The effect of the inter species correlation produces 2-3 mol% change on n-hexane conversion and a second order effect is observed in the di-branched isomer distribution in particular through the thermodynamic factor while the effect of inter species diffusivity is negligible at the investigated operating conditions.

***Keywords*-** **Single-Event MicroKinetics (SEMK), Hydroconversion, ZSM-5, Diffusion, Kinetics**

* Corresponding author . Tel.: +32-(0)9-264-4519; fax: +32-(0)9-264-4999.
E-mail address: Joris.Thybaut@UGent.be (J.W. Thybaut)

Model application to n-hexane hydroisomerization on Pt/H-ZSM-5

Diffusion coefficient as a function of the fractional species occupancy

Figure 2 shows the concentration dependence of the corrected diffusion coefficient, $D_{c,i}$ as a function of the fractional species occupancy [8], for different values of φ, the fraction of acid sites, and λ, the ratio of the average residence time on the acidic and non-acidic sites. In the case of HZSM-5, with an increase in species occupancy, the strong adsorbing sites which are essentially the acid sites, are occupied preferentially. With a further increase in species occupancy, since the sites that interact most strongly with the adsorbing components are no longer available, the diffusion takes place via sites that interact less strongly and, hence, the diffusion is enhanced. With an increase in λ, this phenomenon becomes more prominent. However, when the sorbate concentration goes beyond the total acid site concentration, the molecules start hindering each other at the weak adsorbing sites resulting in a drop of the jump frequency and, hence, the diffusion coefficient. The phenomenon of increasing diffusion coefficient at low adsorbate concentration is also observed experimentally by several researchers [9],[11]. The phenomenological work by Reed and Ehrlich [12] for describing neighbouring atom interaction based on the quasi-chemical approach and its implementation by Krishna et al [13] for ZSM-5 lattice also show a similar pattern for the concentration dependence of the Stefan-Maxwell or the corrected diffusion coefficient, $D_{c,i}$. It can be recalled that in the present work the concentration dependence of the self and the corrected or the Stefan-Maxwell diffusion coefficient is considered identical under the approximations of mean field theory.

The concentration dependence functionality of the diffusion coefficient is developed considering equal jump probability from an intersection site to a channel site located either in the straight or the zigzag channel (see ***Figure 1***) and vice-versa which may not be the most realistic at the hydroisomerization operating conditions. For a rough estimation of the effects of the types of channel sites, an extreme case is simulated which assumes that the activated jump of the physisorbed molecules from an intersection site is only possible to the straight channels while excluding any jump involving the zigzag channels. As a result, 2 channel sites (α) are considered per unit cell instead of 4. Considering a single site per intersection, the following expression is obtained for the self diffusion coefficient of a species *i*.

$$D_{s,i}(\theta) = D_i(\theta \rightarrow 0)[\varphi\lambda_i + 2 - \varphi]\frac{[\varphi\theta_{\beta s} + \lambda_i(1-\varphi)\theta_{\beta w}](1-\theta_{\alpha})}{\lambda_i[(1-\varphi)\theta_{\beta w} + \varphi\theta_{\beta s} + \theta_{\alpha}]} \qquad (1)$$

The Eq. (1) differs from the one proposed by Coppens and Iyenger [8] in terms of the number of channel sites used in the denominator, which decreases by a factor of two. The calculated values of θ_{α}, $\theta_{\beta s}$ and $\theta_{\beta w}$ are also different due to the difference in the number of channel sites.

On the other hand, a cubic lattice containing acid sites offers higher connectivity compared to the ZSM-5 lattice for the activated jump of the adsorbed molecules where only two types of sites can be encountered, strong and weak adsorbing sites. These sites have fractional occupancies, θ_s and θ_w and the concentration dependence functionality of the diffusion coefficient is expressed as follows [4]:

$$D_{s,i}(\theta) = D_i(\theta \rightarrow 0)\frac{[1+\varphi(\lambda_i - 1)](1-\theta\ )}{\varphi(1-\theta_s)(\lambda_i - 1) + (1-\theta)} \qquad (2)$$

Figure 3 compares the normalized corrected diffusion coefficient as a function of fractional species occupancy described. With a decrease in pore connectivity, as in the case of ZSM-5 lattice compared to the cubic lattice, the number of neighbouring vacant sites available for the activated jump decreases. Hence, in presence of the acid sites, the increase in the jump frequency of the molecules at low sorbate concentration by creation of channels of the vacant non-acidic sites is more pronounced in case of the zeolite lattice with low connectivity. As a result, in absence of the jump correlations, a more prominent increase in the diffusion coefficient should be observed for lattice structures with reduced connectivity, as shown in *Figure 3* for the ZSM-5 lattice in comparison to the cubic lattice.

However, with a further reduction in the number of available channel sites (α) per intersection from 4 to 2, the initial increase in the diffusion coefficient is less prominent, contrary to the above explanation. With a reduction in the number of available channel sites (α), the ZSM-5 lattice reduces to a one-dimensional pore structure containing parallel pores with alternate straight and intersection sites. Hence, the molecules cannot jump through creation of alternate pathway of vacant non-acidic sites and the hindrance in the activated jump by the presence of sorbate molecules at the acid sites becomes already prominent at a low occupancy. In a one-dimensional pore structure, since the physisorbed molecules cannot pass each other, anomalous diffusion behaviour is observed leading to deviation from the Einstein's theory of random walk [1]. The observed diffusion phenomenon for the one-dimensional pore structure is another manifestation of this anomalous diffusion behaviour. The higher maximum value obtained in case of the one-dimensional lattice structure may also be related to the molecular motions inside such structure, which takes place in a single file in an ordered way and is less obstructed compared to the random movement of the molecules in the ZSM-5 lattice.

As the concentration dependence of the diffusion coefficient can be related to the catalyst properties, viz. the acid site concentration and the pore connectivity through the model parameters, the effect of these catalyst properties on the conversion and selectivity in n-hexane hydroisomerization is further investigated.

Conversion and product distribution on crystallite and reactor scale

A typical concentration profile inside the ZSM-5 crystallite is shown in *Figure 4* for nC_6, 2MP and 2,3DMB. The diffusion and reaction time scales are comparable, and hence, a concentration gradient is obtained inside the crystallite vide *Figure 4*. The concentration gradient is most pronounced at the external surface of the crystallite [ξ=1] and decreases towards the centre of the crystallite [ξ=0] along the diffusion length. It is to be noted here that the intracrystalline concentration profiles shown in this work correspond to the calculated effluent gas phase composition which is uniform over the entire reactor since a CSTR is used. Hydroisomerization of nC_6 primarily produces mono-branched isomers that diffuse more slowly than n-hexane. The lower diffusivity of the isomers results in an enhanced concentration gradient for these compounds inside the crystallite compared to what would have been obtained in case of equal diffusion coefficients with the feed. Due to the slow diffusion of the mono-branched alkanes inside the crystallite, these undergo further isomerisation reactions. The di-branched products diffuse even more slowly than the mono-branched isomers and are practically almost immobile within the zeolite crystal as evident by their near zero concentration at the crystal surface [ξ=1]. *Figure 4* also depicts the effect of particle shape factor at the crystallite scale. For the same crystallite volume, the diffusion length for the spherical particle is more than that of other geometries, so the effect of component diffusion on concentration profile is more pronounced for spherical particle, as expected [10].

The ratio of the fractional species occupancies of the mono-branched alkanes to nC_6 increases while going from the external surface towards the centre of the crystallite but is far from reaching the equilibrium value as shown in *Figure 5*. However, the ratio of the individual mono-branched isomer concentrations is maintained at the equilibrium composition inside the crystallite. *Figure 5* also shows that the di-branched isomers are produced till a quasi-equilibrium is established between the mono and di-branched isomers towards the centre of the crystallite. In general the branched alkanes are susceptible to β-scission reactions, however for nC_6 the only isomer that can crack through a (s,s) β-scission is 2-methyl-pentane. As a result, cracked product formation is not significant at the operating condition used in this simulation.

As evident from the above discussion, the intracrystalline product profiles are governed by the relative values of the component diffusion coefficients. The various interactions inside the crystallite in terms of the concentration dependence of the diffusion coefficients and the inter-species correlation become important at high species occupancies. However, the effects of correlations in component diffusion at different operating conditions are only visible when there is a change in fractional species occupancy with a change in operating conditions, e.g., at different molar hydrogen to hydrocarbon ratios keeping other operating conditions fixed. It is therefore important to properly design the experiments to study various correlation effects on diffusion and their effect on the product distribution. In the present simulation, where the total occupancy is around 0.8, significant effects are obtained to the product distribution as will be discussed in the subsequent sections.

Similar to the crystallite scale, the product distribution at the reactor scale is also governed by the relative orders of magnitude of the diffusion coefficients for the different compounds as illustrated in *Table 2*. Both the nC_6 conversion and DMB distribution increases if the diffusion coefficient values used in the model are set equal to that of the fastest diffusing component, i.e. nC_6. In the latter case, the products diffuse at a faster rate and the concentration gradient of the products inside the crystallite decreases significantly compared to that of the base case values reported in *Table 1*. As a result, an increase in the conversion and product distribution is observed. The effects of a change in the diffusion coefficients of the fastest and the slowest components, nC_6 and the di-branched isomers on the product distribution is also shown in *Table 2*. With a decrease in the diffusion coefficient for nC_6 keeping the other values fixed at the reference case, the conversion decreases due to a decrease in the yield of mono-branched isomers. With a similar decrease in the diffusion coefficient of the di-branched isomers, the change is conversion is less pronounced compared to the former case but the di-branched isomers distribution decreases significantly.

Confinement effects in presence of lattice heterogeneity

Effect of acid sites concentration

Figure 6 shows the nC_6 conversion as a function of φ, i.e., the acid sites concentration. In the present work, φ affects the diffusion coefficient through concentration dependence functionality. For the purpose of comparison, the dependence of nC_6 conversion on φ is also plotted in *Figure 6* for two other functionalities of the diffusion coefficients: i) a concentration dependence in the form of (1-θ) and ii) no dependence on concentration. The value of λ for n-hexane and its isomers are fixed in this simulation to the calculated value of 14. Considering the concentration dependence in the form of (1-θ), the nC_6 conversion results in a substantial decrease in conversion compared to the other two scenarios. The differences corresponding to these cases are more prominent at a high value of φ and, hence, at high conversion. The decrease in the reaction time scale with an increase in acid site concentration at a given space time is obvious. In presence of the concentration dependence of the diffusion coefficient, the diffusion time scale increases significantly leading to a pronounced decrease in the n-hexane conversion with respect to the case where diffusion coefficient is independent of species concentration. With the functionality for the concentration dependence used in the present work, the values of the normalized diffusion coefficients are always higher compared to that with the linear functionality (1- θ) except at the limits of θ= 0 and 1, cf. Figure 2. At the total species occupancy (θ= 0.7) and λ, used in the present simulation, the normalized diffusion coefficient gives a lower value compared to the case of no dependence, i.e. $D_{s,i}(\theta)/D_i\ (\theta \to 0) = 1$ for all values of φ. Hence, the nC_6 conversion corresponding to the present model lies in between the two limiting cases i) and ii) described above. In case of nC_6 hydroisomerization, as the effect of φ is uniform in the concentration dependence function for the diffusion coefficients of the feed and the product components, no change in the product selectivity is observed (not shown). However, for higher alkanes, the cracking becomes important and produces components with different carbon numbers. In the latter case, the components having differences in their physisorption enthalpies lead to a variation in the λ *[8]*, which may result in a difference in the product selectivity.

Effect of pore connectivity

The effect of the pore connectivity on the concentration profile in the zeolite crystallite is shown in *Figure 7.* The ZSM-5 lattice and the cubic lattice structures are compared. The ZSM-5 lattice structure as shown in ***Figure 1*** represents the MFI pore topology while the cubic lattice can be assumed to represent the FAU pore topology which consists of supercages containing acidic and non-acidic sites. For the concentration dependence functionalities of the diffusion coefficient, the equation proposed by Coppens and Iyenger [8] and Eq. (2) are used respectively for the ZSM-5 lattice and the cubic lattice. For the same value of φ and λ , the diffusion coefficient at a negligible species concentration is higher in case of the cubic lattice owing to its higher pore connectivity. However, the initial increase in the diffusion coefficient as a function of the fractional species occupancy is more pronounced for ZSM-5 compared to the cubic lattice in absence of any jump correlations, leading to a lower maximum in the normalized diffusion coefficient for the latter, cf. *Figure 3.* At the value of the total species occupancy in the present simulation, the decreasing trend in the diffusion coefficient is prominent and the value of the single component corrected diffusion coefficients for ZSM-5 lattice is lower than that of the cubic lattice. Accordingly, the effect of component diffusion on concentration profile is more pronounced for ZSM-5 lattice as shown in *Figure 7.* The corresponding nC_6 conversion is 1.4 mol% lower than for the cubic lattice while the differences in the isomer selectivity were found not to be significantly different.

Effect of drag induced correlation

The correlation between diffusion coefficients of different species is incorporated via the inter-species diffusivity D_{ij} through the inverse of the drag coefficient matrix, $[B]^{-1}$ as well as by the matrix of the thermodynamic factor $[\Gamma]$. In absence of the correlation effects, the component diffusivities are independent of each other and the matrix $[B]^{-1}[\Gamma]$ takes the form of diagonal matrix in terms of the so-called "Fickian" diffusivities. In the latter case, the diffusion coefficient of nC_6 along the crystallite length is not influenced by the diffusion of other components in the mixture. Due to a higher diffusion coefficient of nC_6, at steady state, the occupancy by nC_6 inside the crystallite does not change appreciably from the occupancy at the external surface as shown in *Figure 8*. Owing to higher concentration of nC_6, an increase in the concentration gradient of the product species, e.g., 2MP inside the crystallite is observed.

The effect of the drag induced correlation is more pronounced for a feed mixture compared to a feed of pure nC_6 as shown in *Figure 9* for a mixture of nC_6 and 3MP in equal molar ratio. For a feed mixture, the fractional occupancies of different species are significant at the crystallite surface and have a substantial effect on the inter-species correlation through its constituent matrices $[B]^{-1}$ and $[\Gamma]$. The correlation between species diffusivities is found to be dominated by the matrix of the thermodynamic factor $[\Gamma]$ as the concentration profiles are almost unchanged when only the inter species diffusivity, but not the thermodynamic factor, is removed from the model, see *Figure 9*. The fast moving nC_6 is slowed down in presence of the less mobile branched isomers leading to an increase of the extent of reaction. As a result, the rate of product formation is increased. The effect of the "slowing down" of the more mobile species is more than that of the "speeding up" the slow ones [2]. As a result, the diffusion coefficient of the 2MP does not increase appreciably leading to a larger concentration gradient inside the crystallite.

The effect of inter species correlation on the reactor scale is shown in *Table 3*. As the net rate of product formation increases inside the crystallite, the nC_6 conversion at the reactor outlet increases when the inter species correlation is neglected, i.e. $[\Gamma]$ is considered a unity matrix along with $D_{ij} \rightarrow \infty$. The increase in the di-branched isomer formation is more prominent compared to that of the mono-branched isomers when the drag induced correlation is neglected and manifested in almost two-fold increase in the DMB distribution.

Conclusions

The single-event microkinetic methodology has been integrated for the first time with a rigorous modelling of the diffusion phenomena. The diffusion phenomenon at the molecular scale is modeled by considering the interactions with the pore wall and between the species themselves. Confinement effects for the multicomponent mixtures in the presence of the lattice heterogeneity are translated to the crystallite scale and solved simultaneously with the reaction kinetics to obtain the concentration profile in the crystallite and the reactor outlet composition.

The model accounts explicitly for the pore structure by distinguishing between physisorption and diffusion of the linear and the branched alkanes contributing to the reactant and product shape selectivity. Significant effects on conversion are observed through the concentration dependence of the diffusion coefficient as well as the drag induced correlation. The latter also leads to a change in product selectivity though the effects are in second order in particular via the thermodynamic factor while the effect of the inter-species diffusivity is negligible at the investigated operating range.

The developed model accounts explicitly for the concentration of acid sites and distinguishes between the latter and the other physisorption sites. The pore connectivity is accounted for via the concentration dependence of the diffusion coefficients. These catalyst descriptors allow to describe quantitatively the effects of catalyst properties on the intracrystalline concentration profiles as well as on the overall conversion and product distribution.

Hence, the model can be used for further understanding of the sorption and diffusion induced shape selectivity in hydroconversion on ZSM-5 zeolites and other medium pore zeolites having similar structure.

Acknowledgements

This research has been carried out as a part of the FS2 (Functional Supramolecular Systems) in the framework of the Interuniversity Attraction Poles Programme funded by the Belgian Science Policy.

References:

[1] J. Kärger, D.M. Ruthven, Diffusion in Zeolites and Other Microporous Solids, John Wiley, New York, 1992.

[2] R. Krishna, J.A. Wesselingh, Chem. Eng. Sci. 52 (1997) 861-911.

[3] M-O. Coppens, A.T. Bell, A.K. Chakraborty, Chem. Eng. Sci. 53 (1998) 2053-2061.

[4] M-O. Coppens, A.T. Bell, A.K. Chakraborty, Chem. Eng. Sci. 54 (1999) 3455-3463.

[5] V. Iyenger, M-O. Coppens, Chem. Eng. Sci. 59 (2004) 4747-4753.

[6] J.W. Thybaut, G.B. Marin, G.V. Baron, P.A. Jacobs, J.A. Martens, J. Catal. 202 (2001) 324-339.

[7] J.W. Thybaut, C.S.L. Narasimhan, G.B. Marin, J.F.M. Denayer, G.V. Baron, P.A. Jacobs, J.A. Martens, Catal. Lett. 94 (2004) 81-88.

[8] M-O. Coppens, V. Iyenger, Nanotechnology 16 (2005) S442-S448.

[9] A.O. Koriabkina, A.M. de Jong, E.J.M. Hensen, R.A. van Santen, Micropor. Mesopor. Mater. 77 (2005) 119-129.

[10] W. Zhu, A. Malekian, M. Eic, F. Kapteijn, J.A. Moulijn, Chem. Eng. Sci. 59 (2004) 3827-3835.

[11] L. Song, L.V.C. Rees, J. Chem. Soc. Faraday Trans. 93 (1997) 649-657.

[12] D.A. Reed, G. Ehrlich, Surf. Sci. 102 (1981) 588-609.

[13] R. Krishna, D. Paschek, R. Baur, Micropor. Mesopor. Mater. 76 (2004) 233-246.

List of Figures

parameters taken from Thybaut et al., 2004 [7]; Full line: DMB/MP, Short Dashed line: MP/nC_6, Long dashed line: 3MP/2MP; Thick line: model calculation, Thin line: equilibrium composition; crystallite dimension =15 μm, slab geometry Figure 6Figure 6: nC_6 conversion as a function of φ, calculated by the SEMK model at T = 503 K, P = 1.0 MPa, inlet molar H_2/ nC_6 ratio = 50 and W/F^0_{HC}: 134 kg-s mol^{-1} using the diffusion coefficients given in *Table 1* and other model parameters taken from Thybaut et al., 2004 [7]; Full line: model calculation with concentration dependence of the diffusion coefficients [8] at λ = 14, Long dashed line:, model calculation with no concentration dependence of the diffusion coefficients, Short dashed line: model calculation with concentration dependence of the diffusion coefficients in the form of (1-θ); crystallite dimension =15 μm , spherical geometry

Figure 7: Concentration profile as a function of dimensionless length of the zeolite crystallite for ZSM-5 lattice and cubic lattice structures calculated by the SEMK model at T = 503 K, P = 2.0 MPa, inlet molar H_2/HC ratio = 50 and W/F^0_{HC}: 134 kg-s mol^{-1} using the diffusion coefficients given in *Table 1*, φ = 0.2 and λ = 14; Full lines: model calculation for ZSM-5 lattice with concentration dependence function [8]. Dashed lines: model calculation for cubic lattice with concentration dependence function vide Eq. (2) [4]; Thick lines: nC_6, Thin lines: 2MP; crystallite dimension =15 μm , spherical geometry

Figure 8: Concentration profile as a function of dimensionless length of the ZSM-5 crystallite describing the effect of correlation between component diffusivities calculated by the SEMK model at T = 503 K, P = 1.0 MPa, inlet molar H_2/nC_6 ratio = 50 and W/F^0_{HC}: 134 kg-s mol^{-1} using the diffusion coefficients given in *Table 1*; Full line: present model, Dashed line: in absence of the correlation between component diffusivities; Thick lines: nC_6, Thin lines: 2MP; crystallite dimension =15 μm, spherical geometry

Figure 9: Concentration profile as a function of dimensionless length of the ZSM-5 crystallite describing the effect of correlation between component diffusivities calculated by the SEMK model at T = 503 K, P = 1.0 MPa, inlet molar H_2/HC ratio = 50 and W/F^0_{HC}: 95 kg-s mol^{-1}, Feed composition : nC6 : 3MP = 50: 50; full line: present model, long dashed line: in absence of the correlation between component diffusivities, short dashed line: present model in absence of the inter species diffusivity, i.e. $D_{ij} \rightarrow \infty$; thick lines: nC_6, thin lines: 2MP; crystallite dimension =15 μm, spherical geometry

Figure 1

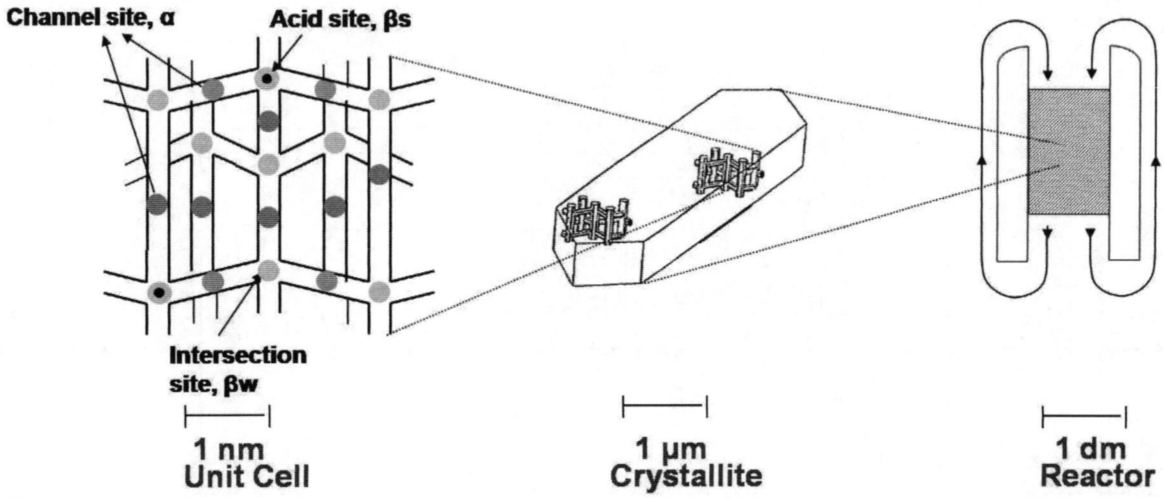

Channel site, α
Acid site, βs
Intersection site, βw
1 nm
Unit Cell
1 μm
Crystallite
1 dm
Reactor

Figure 2

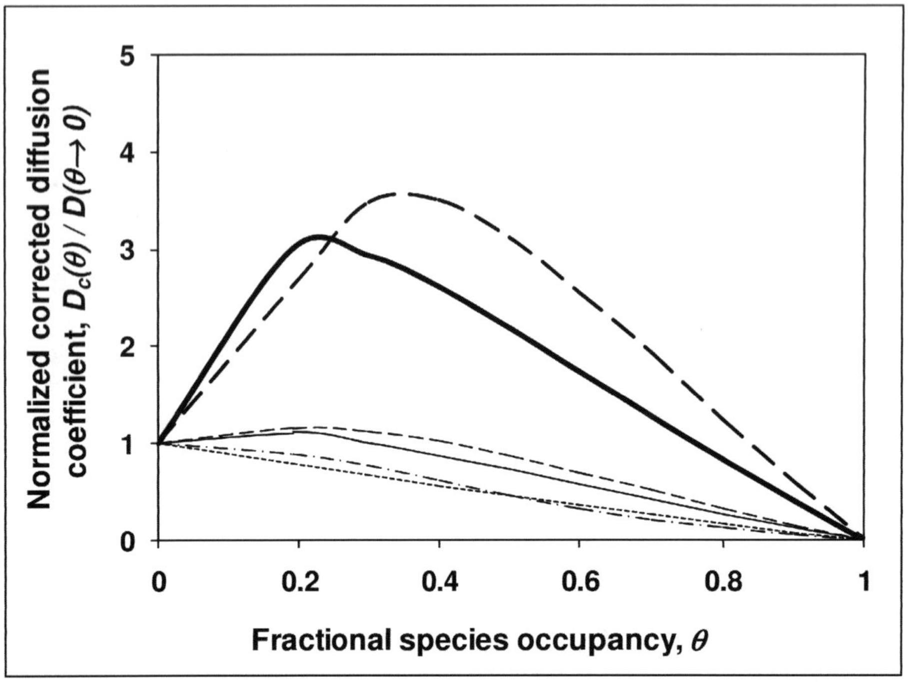

Normalized corrected diffusion coefficient, $D_c(\theta) / D(\theta \to 0)$
5
4
3
2
1
0
0
0.2
0.4
0.6
0.8
1
Fractional species occupancy, θ

Figure 3

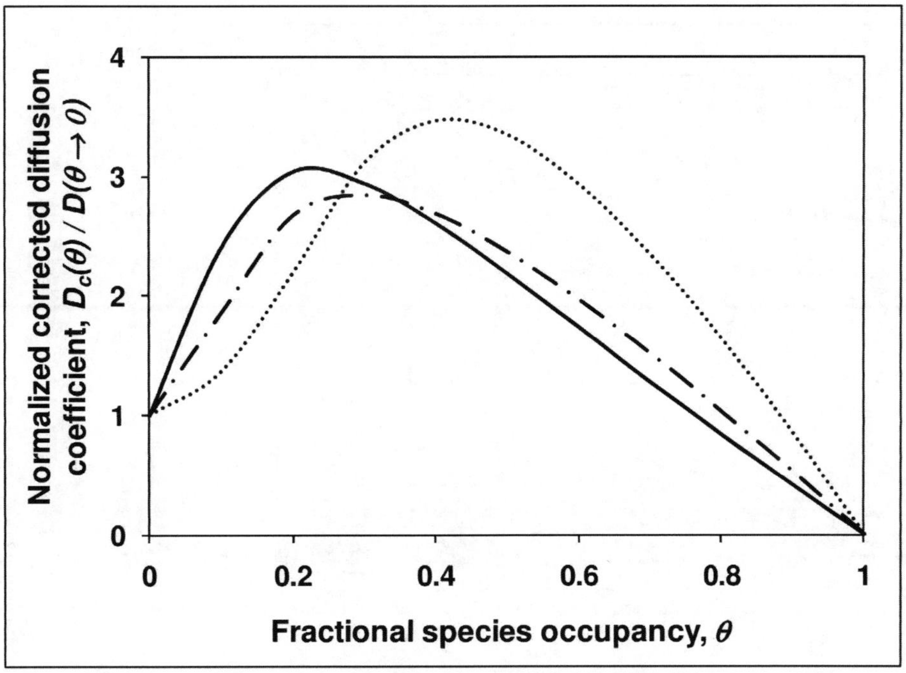

Normalized corrected diffusion coefficient, $D_c(\theta) / D(\theta \to 0)$
4
3
2
1
0
0
0.2
0.4
0.6
0.8
1
Fractional species occupancy, θ

Figure 4

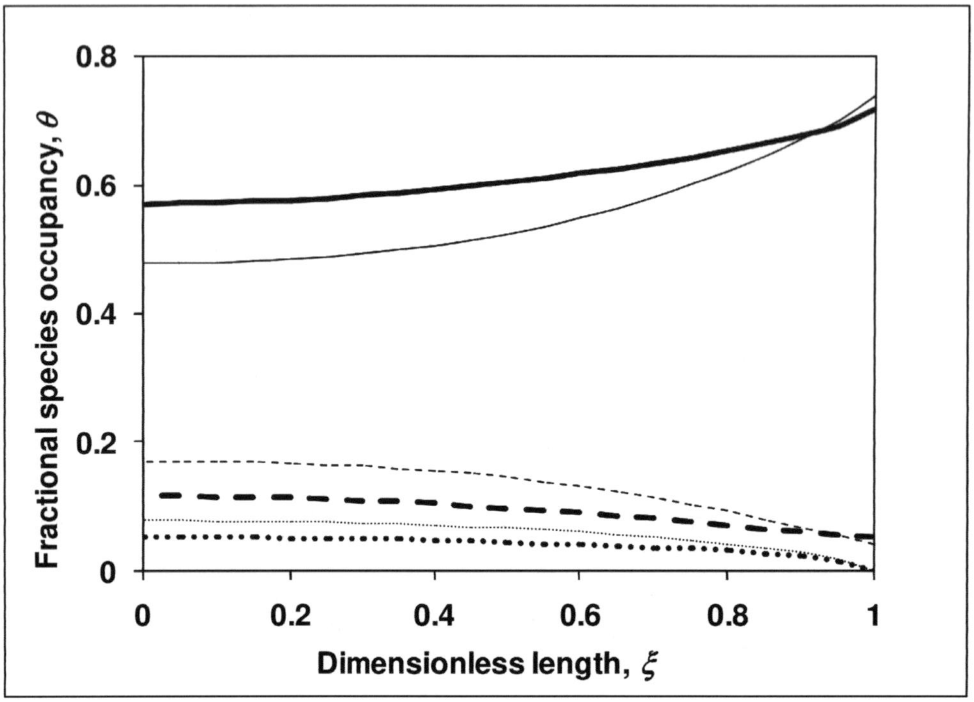

0.8
0.6
0.4
0.2
0
Fractional species occupancy, θ
0
0.2
0.4
0.6
0.8
1
Dimensionless length, ξ

Figure 5

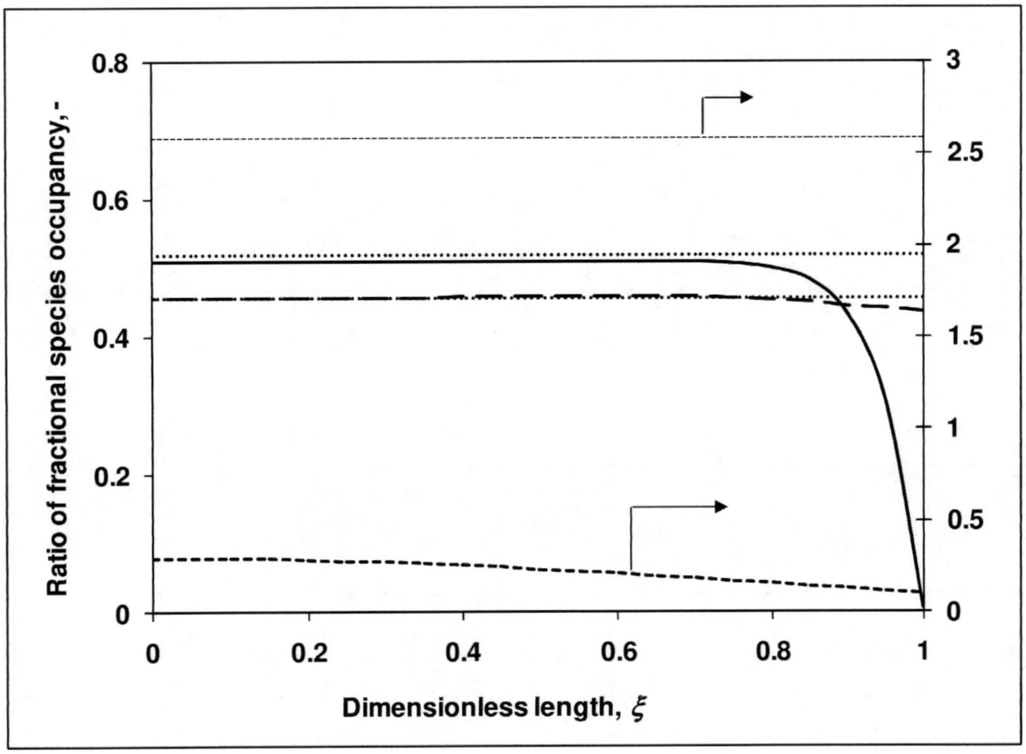

0.8
0.6
0.4
0.2
0
3
2.5
2
1.5
1
0.5
0
0
0.2
0.4
0.6
0.8
1
Ratio of fractional species occupancy, -
Dimensionless length, ξ

Figure 6

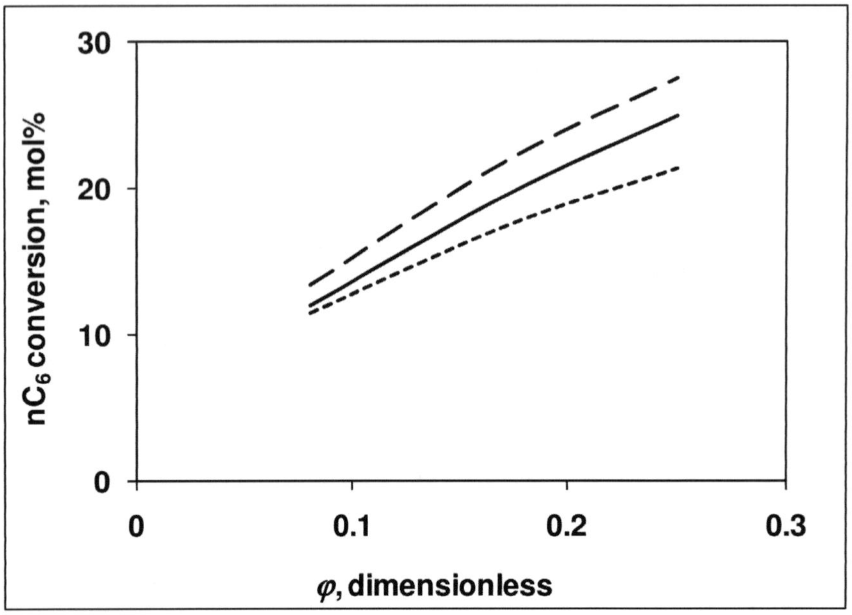

Figure 7

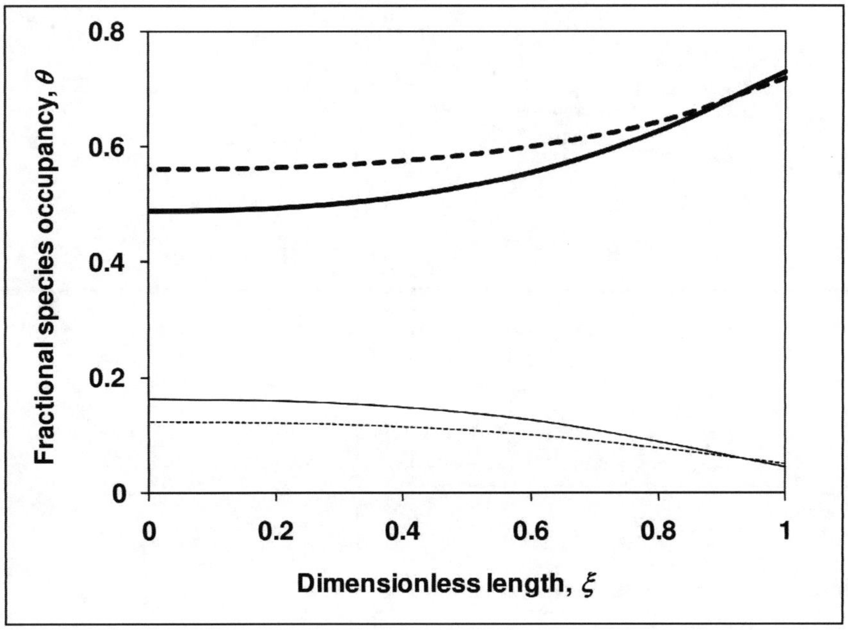

0.8
0.6
0.4
0.2
0
Fractional species occupancy, θ
0
0.2
0.4
0.6
0.8
1
Dimensionless length, ξ

Figure 8

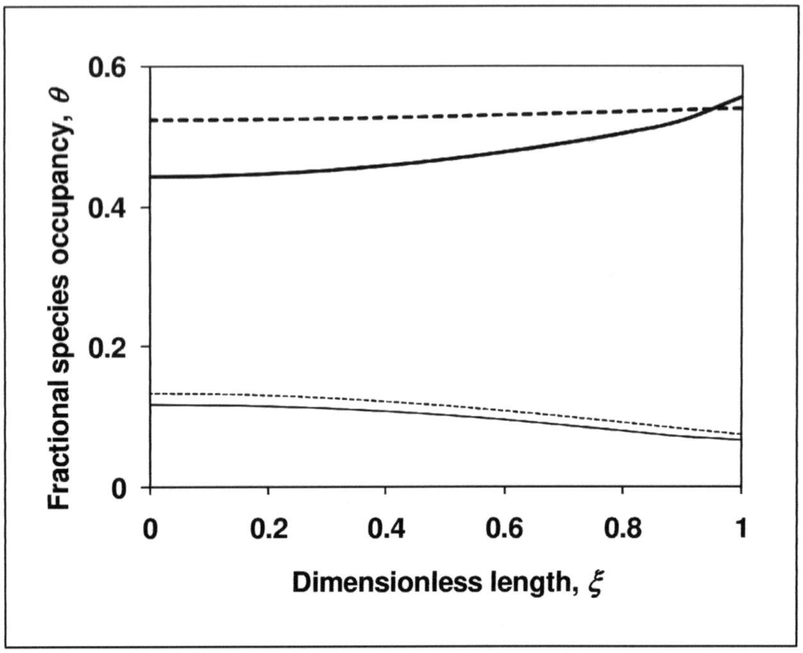

0.6
0.4
0.2
0
Fractional species occupancy, θ
0
0.2
0.4
0.6
0.8
1
Dimensionless length, ξ

Figure 9

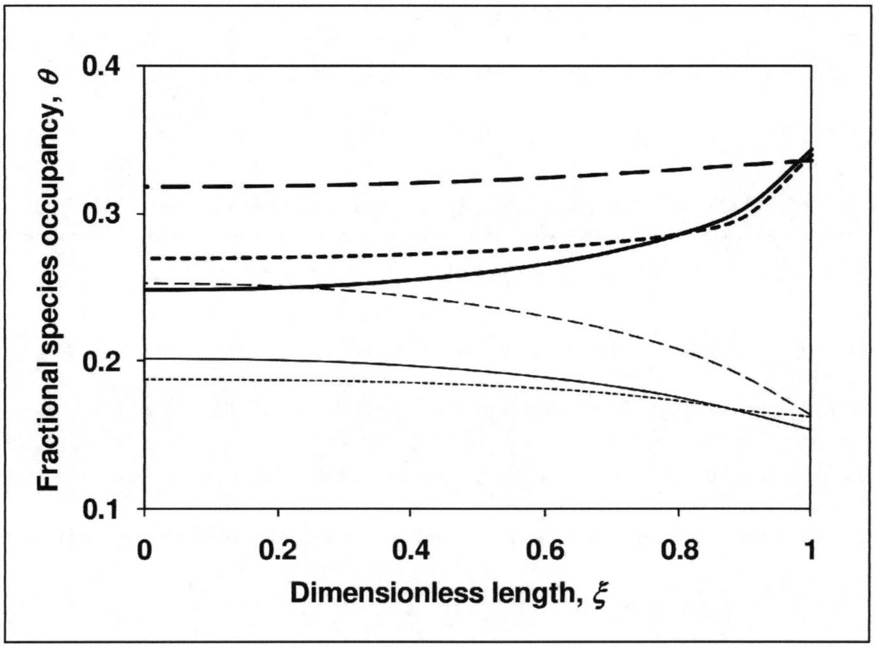

0.4
0.3
0.2
0.1
Fractional species occupancy, θ
0
0.2
0.4
0.6
0.8
1
Dimensionless length, ξ

LIST OF TABLES

Table 1

	nC_6	2MP	3MP	DMB	C_3
$D_i(\theta \rightarrow 0)$ ($10^{-12}\,m^2s^{-1}$) [a]	22	4.5	4.0	2.3 10^{-3}	5.0 10^{3}
E_A (kJ mol^{-1}) [b]	19	29	44	-	-

[a] value at 503 K

[b] $D_i(\theta \rightarrow 0) = D_i^0(\theta \rightarrow 0) * \exp\left(\frac{-E_A}{RT}\right)$

Table 2

$D_{nC_6} / D_{nC_6}^{ref}$ [a]	$D_{DMB} / D_{DMB}^{ref\ a}$ [a]	Conversion	DMB distribution[c]
1	1	14.8	1.3
1	~ 10^4 [b]	17.2	7.6
0.2	1	13.9	1.4
1	0.2	14.6	0.8

[a] Reference case corresponds to the diffusion coefficients vide *Table 1*;

diffusion coefficient of other components are also at the reference case

[b] The diffusion coefficients of other components are equal to that of the n-hexane

[c] $\text{Di-Methyl-Butane distribution} = \frac{F_{DMB}}{F_{Isom}} \times 100$

Table 3

	Conversion	DMB distribution[a]
In presence of inter species correlation	12.2	0.7
No inter species correlation, $[\Gamma]=[I]$, $D_{ij} \to \infty$	14.5	1.3

[a] $\text{Di-Methyl-Butane distribution} = \frac{F_{DMB}}{F_{Isom}} \times 100$

Examine the Diffusive Behavior of Different Combinations of Ethanol and Water within An Amorphous Mixture of PDMS and High Silica ZSM-5 Zeolite by Molecular Simulation

Paul F. Harten, Sustainable Technology Division, U.S. Environmental Protection Agency, Cincinnati, OH

An amorphous mixture of PDMS and multi-cellular fragments of ZSM-5 are brought together to approximate the properties of a mixed matrix membrane of PDMS with ZSM-5. The diffusive behavior of different ethanol and water combinations within this amorphous mixture are simulated by NVT-ensemble equilibrium molecular dynamics techniques. The effect of concentration on the diffusive behavior of the components is investigated through the analysis of both mean square displacements and radial distribution functions. Agreement with experimental work suggests the similar simulation of many different amorphous mixtures by computer to help identify other promising mixed matrix membranes which may be used to facilitate the separation of ethanol from water.

See more of this Session: Industrial Applications of Computational Chemistry and Molecular Simulation II
See more of this Group/Topical: Computational Molecular Science and Engineering Forum

The Fourier Correlation Method for Predicting Diffusivities in Multicomponent Mixtures Simulated by Molecular Dynamics

Dean R. Wheeler, Chemical Engineering, Brigham Young University, Provo, UT

Joseph W. Nichols, Chemical Engineering, Brigham Young University, Provo, UT

As has been recognized for several decades, molecular dynamics simulations can be an elegant way to calculate and predict transport properties, including diffusion coefficients. The active development in the field over the years has led to a number of simulation techniques for diffusivity, most notably the Green-Kubo relation for equilibrium molecular dynamics as well as nonequilibrium molecular dynamics (NEMD) techniques. In this work we report on a new method for predicting bulk diffusivities in multicomponent mixtures. The technique, which we call Fourier correlation (FC), can be implemented through a simple algorithm within existing equilibrium MD computer simulations. The FC method conceptually combines macroscopic equations of change, finite Fourier transforms, and microscopic fluctuations. We briefly outline the rationale and advantages of the method and show preliminary results for obtaining diffusivities in nonideal binary and ternary liquid mixtures.

See more of this Session: Industrial Applications of Computational Chemistry and Molecular Simulation II
See more of this Group/Topical: Computational Molecular Science and Engineering Forum

Prediction of Environmental Impact of Energetic Materials *in Silico*

Nandhini Sokkalingam, Department of Chemical Engineering, Detroit, MI

Jeffery J. Potoff, Chemical Enginering, Wayne State University, Detroit, MI

The objective of this research is to determine how some energetic materials behave in the environment. The fate of any compound in soil, water or the atmosphere can be determined by predicting the partition or distribution coefficient. Two key partition coefficients used by the scientific community to assess a compound's impact in the environment are octanol-water partition coefficients and Henry's law constants (air-water partition coefficient). A wide variety of experimental techniques exist to measure the partition coefficient but atomistic simulation is attractive as a predictive tool for studying the environmental impact of these compounds due to three primary reasons: the hazardous nature of the explosives, long experimental time scales involved in testing of the materials and the costs associated with it. Furthermore, compounds can be evaluated for environmental hazards via simulation before synthesis, potentially leading to significant cost savings.

In this work force fields (molecular models) have been developed for six energetic compounds, 2,4-dinitroanisole (DNAN), N-methyl-p-nitroaniline (MNA), dinitropyrazole (DNP), nitrotriazolone (NTO), methyl-trinitroimidazole (MTNI) and TATB (trinitrotriaminobenzene) categorized by the army as potential insensitive munitions compounds. These force fields are used to predict pharmacokinetic properties such as octanol-water partition coefficients and Henry's law constants along with other thermophysical properties such as vapor-liquid equilibria, vapor pressure, critical parameters and normal boiling points. The validation of the developed force fields is accomplished with the scarce experimental data available. The octanol-water partition coefficients and Henry's law constant predicted for DNAN and MNA are in good agreement with the experiment. For DNP, NTO, MTNI and TATB, materials for which limited experimental data are available, the predicted partition coefficients agree well with values calculated with group contribution methods.

See more of this Session: Industrial Applications of Computational Chemistry and Molecular Simulation II
See more of this Group/Topical: Computational Molecular Science and Engineering Forum

A Complete Multiscale Modeling Approach for Polymer-Clay Nanocomposites

Maurizio Fermeglia, Molecular Simulation Engineering Laboratory, Department of Chemical Engineering, University of Trieste, Trieste, Italy

Giulio Scocchi, Icimsi, SUPSI, Manno, Switzerland

Sabrina Pricl, Molecular SImulation Engineering (MOSE) Laboratory - DICAMP, University of Trieste, Trieste, Italy

Jan-Willem Handgraaf, CULGI BV, Leiden, Netherlands

Johannes Fraaije, Leiden Institute of Chemistry, Soft Matter Chemistry, Gorlaeus Laboratories, University of Leiden, Leiden, Netherlands

Paola Posocco, University of Trieste, Trieste, Italy

In order to develop new materials and composites with designed original properties, it is essential for these properties to be predicted before preparation, processing, and experimental characterization. Computer-assisted molecular simulations can then play a major role in this context. Given these concepts, however, it is necessary to carry out calculations for realistic time scales fast enough to be useful in design. This requires developing techniques useful to design engineers, by incorporating the methods and results of classical, low scales quantum mechanics/molecular dynamics (QM/MD) simulations to mesoscale modeling (MS) to finite element (FE) calculations [1-3]. Here, we report the results of our efforts in developing a hierarchical procedure for bridging the gap between atomistic and MS/FE simulations for PCN design. According to the computational recipe, all necessary parameters of the mesoscopic models are estimated by an ad hoc, step-by-step procedure involving different QM/MD calculations. Finally, the mesoscopic simulated structures are passed on to the FEM calculations, to estimate the macroscopic properties of the PCN (e.g., Young modulus, diffusivity, etc.). The global perspective of our work is the complete integration of all available simulation scales, in a hierarchical procedure to provide an efficient and robust simulation protocol for the successful design of PCNs of industrial interest and the prediction of their final performance.

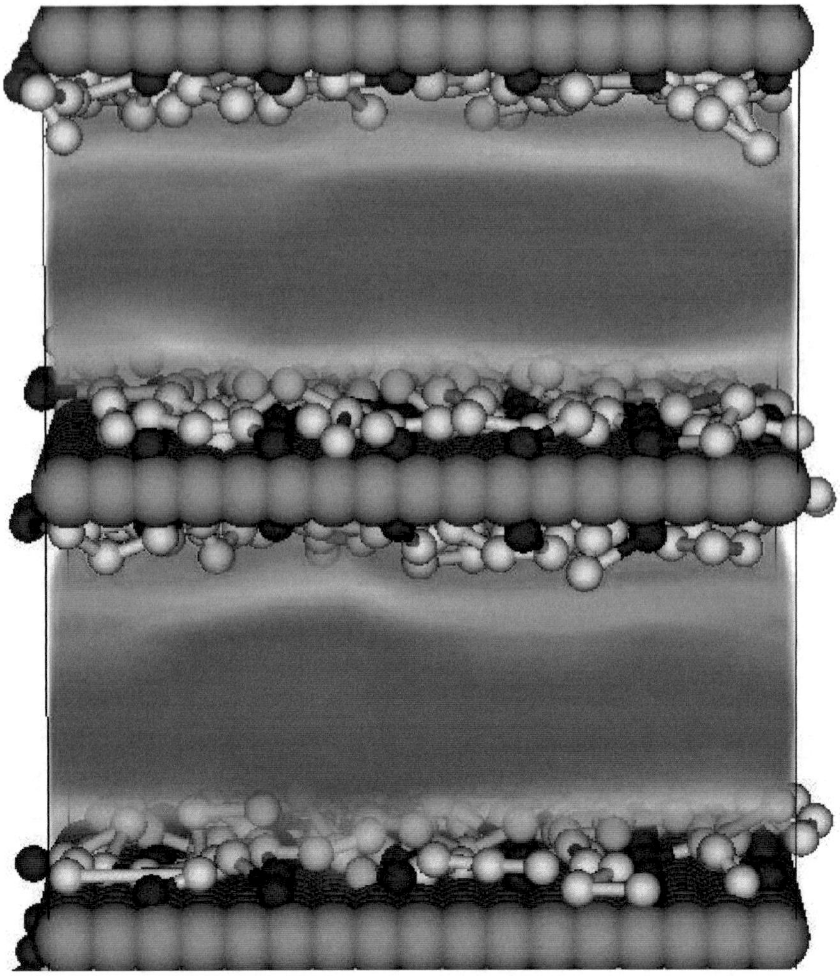

See more of this Session: Industrial Applications of Computational Chemistry and Molecular Simulation II

See more of this Group/Topical: Computational Molecular Science and Engineering Forum

Investigation of Epoxy Network Formation by Simulation and Experiment

Michael Aldridge, Department of Materials Science & Engineering, University of Michigan, Ann Arbor, MI

Chandrashekar Shankar, Chemical Engineering, University of Pittsburgh, Pittsburgh, PA

Changgua Zhen, Department of Materials Science & Engineering, University of Michigan, Ann Arbor, MI

John Kieffer, Department of Materials Science & Engineering, University of Michigan, Ann Arbor, MI

Diglycidyl ether of bisphenol-A (DGEBA) and diethylenetriamine (DETA) react to form a highly cross-linked epoxy network. We investigated the cure kinetics of this epoxy system as a function of its chemical composition using a combination of inelastic light scattering and molecular dynamics (MD) simulations. Using concurrent Brillouin and Raman scattering we monitored both the changes in visco-elastic properties and chemical signatures of the epoxy in situ, while it was polymerizing. The complex mechanical modulus revealed by Brillouin scattering provides the most direct measure of the structural integrity of the polymer network. In order to establish the relationship between degree of cure and the elastic modulus of the resulting structure, we carried out multi-scale molecular simulations. We generated epoxy networks using a combination of MD and a heuristic bond exchange protocol. Mechanical properties of these networks were evaluated, by subjecting them to tensile and shear tests at various degrees of reaction. Raman active vibrational modes for the various structural moieties were predicted using density functional theory calculations and weighted according to their concentrations in MD-simulated structures to generate Raman spectra as a function of the degree of cure. Finally, all experimental and simulation data were combined in a system of differential equations to establish a model for the epoxy cure kinetics in way consistent with all methods of investigation.

See more of this Session: Industrial Applications of Computational Chemistry and Molecular Simulation II
See more of this Group/Topical: Computational Molecular Science and Engineering Forum

Investigations of Ring Opening Metathesis Polymerization (ROMP) of Cross-Linked Dicyclopentadiene (DCPD) Networks

Chandrashekar Shankar, Chemical Engineering, University of Pittsburgh, Pittsburgh, PA

Changgua Zhen, Materials Science and Engineering, University of Michigan, Ann Arbor, MI

Michael Aldridge, Materials Science and Engineering, University of Michigan, Ann Arbor, MI

John Kieffer, Department of Materials Science & Engineering, University of Michigan, Ann Arbor, MI

Cross-linked DCPD networks are experimentally created by a ROMP mechanism mediated by Ruthenium based Grubb's catalysts, resulting in a wide variety of polymer architectures and ensuing mechanical properties. We generated DCPD networks using a combination of molecular dynamics (MD) and a novel heuristic bond exchange protocol recreating the ROMP pathway during the curing process. The reaction scheme is guided by structural rearrangement features verified via Raman spectroscopy and first principles calculations on intermediate moieties in the reaction pathway. Mechanical properties of the simulated networks at various degrees of reaction were evaluated by subjecting them to uniaxial tension tests to calculate the Young's modulus and by isotropic compression to calculate the bulk modulus. Experimental validation of modulus values is achieved via in-situ Brillion spectroscopy during the curing process. The results suggest much promise for our novel algorithm for other network systems such as epoxies.

See more of this Session: Industrial Applications of Computational Chemistry and Molecular Simulation II
See more of this Group/Topical: Computational Molecular Science and Engineering Forum

Interactions of Epoxy-Based Polymers with Carbon Nanotubes Studied by Molecular Modeling

Andreas Bick, Scienomics SARL, Dortmund, France

Klaus B. Stark, Scienomics SARL, Dortmund, France

In this study initially molecular models of of epoxy-based cross linked polymers are built and investigated by molecular dynamics simulation. Properties like density, solubility parameters and elastic moduli are computed and compared against experimental results and also an idealized linear epoxy system. Then a mixed system of single walled carbon nanotubes and the polymer matrix is investigated, analyzing the impact of the CNT filler on the properties of the system. Finally the results from the atomistic simulations are used to parameterize a mesoscale Dissipative Particle Dynamics simulation (DPD) to investigate the impact of cross linking and chemical detail on the distribution of the carbon nanotubes in the polymer matrix.

See more of this Session: Industrial Applications of Computational Chemistry and Molecular Simulation II
See more of this Group/Topical: Computational Molecular Science and Engineering Forum

Design of Polymer Scaffolds for Supported Catalysts Using Molecular Simulations

Andrew Swann, School of Chemical & Biomolecular Engineering, Georgia Institute of Technology, Atlanta, GA

Wei Zhang, School of Chemical & Biomolecular Engineering, Georgia Institute of Technology, Atlanta, GA

Peter J. Ludovice, School of Chemical & Biomolecular Engineering, Georgia Institute of Technology, Atlanta, GA

Marcus Weck, Department of Chemistry, New York University, New York

Polynorbornene produced by Ring Opening Metathesis Polymerization (ROMP) has shown promise as a scaffold to support tethered homogeneous catalytic sites. Such catalyst supports provide the advantage of efficient homogeneous catalytic sites while making it easy to separate such catalysts from the reaction mixture. Molecular simulation has been applied to various isomers of this polymer to determine how the backbone structure affects the conformation of this polymer and consequently the degree to which the tethered catalyst sites are exposed to potential reactions. Previously, this ROMP polymer was assumed to adopt an expanded conformation similar to the vinyl-polymerization version of the same polymer. This assumption was based on the theory that polymers with alternating non-rotatable backbone bonds and bulky side groups formed elongated helical structures. This theory was derived from simulations of vinyl polymerized polynorbonene and substituted polyacetylenes. Simulations of ROMP polynobornene failed to exhibit this elongated helical structures because of the flexibility of the backbone cyclopentyl rings in this polymers. Simulations of ROMP poly(norbornene) in which the flexibility of the backbone cyclopentyl ring was reduced with a fused imide group did show signs of helical conformations, but only when a sufficiently large side chain was attached. Bulk simulations of this modified ROMP polynorbornene predicted the formation of intermediate bulk structure, which is often associated with helical conformations. The Wide-Angle X-Ray Diffraction (WAXD) pattern predicted from these simulations reproduced the experimentally measured one for a synthesized ROMP polynorbornene. These results provide useful heuristics for the design of optimal industrial catalyst platforms.

See more of this Session: Industrial Applications of Computational Chemistry and Molecular Simulation II
See more of this Group/Topical: Computational Molecular Science and Engineering Forum

A Mulitscale Simulation Approach to Issues in Drug Delivery

Simon C. McGrother, Culgi, Raleigh, NC

Johannes Fraaije, Leiden Institute of Chemistry, Soft Matter Chemistry, Gorlaeus Laboratories, University of Leiden, Leiden, Netherlands

Enhanced drug delivery devices offer a number of benefits, including: increased patient compliance, more even dosages, extended patent-life and marketplace differentiation. We report on a series of multiscale simulations of drug delivery from various depot devices. Multiscale simulation involves techniques ranging from atomistically detailed molecular mechanics models, to coarse-grained representations using dissipative particle dynamics and hybrid methods. At the coarse-grained level some detail is sacrificed in the interest of probing longer length- and time-scales. The utility and benefits of the various methods are discussed. In particular we look at release profiles of drugs from polymeric devices as a function of the relative hydrophobicity of the active ingredient and of the nature of the polymeric device. The role of electrostatic interactions is investigated. In addition we carefully consider issues of coarse-graining and reverse-mapping (replacing a coarse-grained model with its fully atomistic equivalent).

See more of this Session: Multi-Scale Modelling II
See more of this Group/Topical: Computational Molecular Science and Engineering Forum

From Molecules to Life: Multiscale Models in Biology

Yiannis N. Kaznessis, Chemical Engineering and Materials Science, University of Minnesota, Minneapolis, MN

We attempt to tackle what many consider the biggest question in the biosciences: how does life emerge from a soup of chemicals? Phrased differently, can biological phenotypes be explained with mathematical models of molecules that interact according to physical laws?

At the crux of the matter lies the doubt that humans can develop faithful mathematical representations of living organisms, doubt fed by the astonishing complexity of biological systems. However intense are the efforts of biologists to reduce organisms to sequences, structures and interactions, the difficulty to develop tractable mathematics hampers the connection between theory and reality in biology.

Why is that? Biological systems are not only non-linear and often stochastic; they possess an overwhelming number of variables. Consequently, although in principle these systems obey physical laws, there are insurmountable mathematical difficulties in developing tractable, predictive models. It is no surprise then that mathematics is not considered as indispensible a tool in the biosciences as in the physical sciences. Besides, biology is a discipline in history: Dobzhansky's dictum that "Nothing in biology makes sense except in the light of evolution" casts a long shadow on mathematical models of phenotypic complexity. Because, how exactly can we integrate thermodynamics with evolution?

We believe that a synthetic biology research programme may liberate empiricism beyond the unaided human brain. Synthetic biological systems confer three advantages: a) they are small and well-defined enough to be captured by universal yet tractable mathematical models; b) they are modular enough to string together and build logical and informational architectures that are the essence of living systems; c) they are our designs, not nature's, avoiding the difficulties of historical explanations.

What is distinctive then is that instead of a top-down approach ("systems biology") we start from the bottom and go up: designing small synthetic biological systems affords the luxury of knowing components and their interactions well enough to develop mechanistic models and test whether reductionism applies to biology. Furthermore, combining these systems together can result in the logical and informational architectures that Nobel Laureate Paul Nurse (Medicine, 2001) considers as the missing link between chemical molecules and the miracle of life. If there were a chance to build reductionist, mechanistic models and capture biological complexity, it is probably with careful modeling of modular synthetic systems. Armed with supercomputers, we can capture synthetic biocomplexity and provide augmented insight invaluable to all biosciences.

We are explaining biological phenotypes in terms of biomolecular interaction networks that obey laws of statistical thermodynamics and principles of molecular biology. We develop mathematical descriptions that span multiple levels of organization from molecules to logical architectures.

Specifically, we are developing statistical mechanical methods that predict the interaction strength between biomolecules. For example, we recently published a manuscript in the JACS, entitled "Path-integral method for predicting relative binding affinities of protein-ligand complexes". It is a holy grail in computational biology to develop accurate and efficient methods for computing binding free energies. We believe our method is at least as accurate as previous methods, while certainly being more efficient.

We are also studying the mathematical rules that dictate the behavior of gene networks. Biological systems are on occasion far from the thermodynamic limit. Stochasticity takes hold then, rendering false traditional methods for modeling reaction kinetics. We are developing relevant mathematical tools. We are focusing on synthetic gene networks, modeling how interactions of biomolecules result in logical and informational architectures, and ultimately in biological phenotypic complexity. We recently published a summary of our software package ("SynBioSS: the Synthetic Biology Modeling Suite." Bioinformatics, 2008,1;24(21):2551-3). SynBioSS is a suite of multiscale algorithms for modeling non-linear, stochastic biomolecular dynamics.

In this presentation, we will discuss different models and how information is transferred from molecules, their interactions, to cascades of gene regulatory relations, to emerging logical and informational architectures in bacteria.

See more of this Session: Multi-Scale Modelling II
See more of this Group/Topical: Computational Molecular Science and Engineering Forum

Modeling the Hydrophobic Effect by Coupling Solutes to a Lattice Gas

Amish J. Patel, Chemistry, University of California, Berkeley, Berkeley, CA

David Chandler, Chemistry, University of California, Berkeley, Berkeley, CA

In problems of biological assembly, manifestation of the hydrophobic effect is complex depending on the size as well as the conformation of the solute. The solute disrupts the inherent H-bonding network of water, thereby causing an unbalancing of attractive interactions experienced by the water molecules in its vicinity. The extent of this disruption determines the solvation free energy of the solute. The theory of the hydrophobic effect proposed by Lum, Chandler and Weeks (LCW) successfully describes this rich interplay between the solute and solvent structures by coarse-graining the solvent density and analytically integrating out solvent fluctuations on length-scales smaller than the coarse-graining length. Since the implementation of LCW theory can be computationally very demanding, the coarse-grained solvent density was mapped onto a lattice gas by ten Wolde, Sun and Chandler. In this work, we further improve upon the theory by relaxing certain assumptions about the unbalancing of attractive interactions on length scales smaller than the coarse-graining length. In addition to a brief overview of the proposed model, the results obtained by application of the model to several pertinent problems of hydrophobic assembly will be presented.

See more of this Session: Multi-Scale Modelling II
See more of this Group/Topical: Computational Molecular Science and Engineering Forum

Molecular Dynamics Simulations of Complex Formation Between Poly(styrene sulfonate) and Poly(allylamine hydrochloride), and Poly(acrylic acid) and Poly(allylamine hydrochloride)

Nazish Hoda, Department of Chemical Engineering, University of Michigan, Ann Arbor, MI

Ronald G. Larson, Chemical Engineering, University of Michigan, Ann Arbor, Ann Arbor, MI

Explicit- and implicit-solvent molecular dynamics simulations are performed to study complexation in two polyelectrolyte systems: poly(styrene sulfonate)/poly(allylamine hydrochloride) (PSS/PAH) and poly(acrylic acid)/poly(allylamine hydrochloride) (PAA/PAH). In both the systems, the polyions have the same number of monomers and are present in stochiometric proportion, and water is used as solvent. Simulations give important insights into the structure and composition of the complexes. We found that the PSS/PAH complex is more compact and has a smaller fraction of water than the PAA/PAH complex due to the presence of phenyl ring in PSS and the hydrophillic nature of the charged group in PAA. The effects of salt concentration and partial charge fraction were also probed. Either an increase in salt concentration or a decrease in partial charge fraction increases complex swelling and the water content in the complex. To overcome the computational limitations associated with explicit-solvent simulations, we performed implicit-solvent simulations. A simple strategy to develop the force field for implicit-solvent simulations is proposed. In the absence of salt and for fully charged polyions, the radius-of-gyration as well as the various radial distribution functions predicted by the implicit-solvent simulations are in excellent agreement with the explicit-solvent simulations, while reasonable agreement is obtained in the other cases. The implicit-solvent simulations are performed for bigger, realistic systems, and we found that the general trends observed in these simulations agree with those for smaller system sizes, suggesting that the explicit-solvent simulation results are independent of system size. Finally, the findings of the simulations are used to explain the origin of exponential growth in layer-by-layer assembly.

See more of this Session: Multi-Scale Modelling II
See more of this Group/Topical: Computational Molecular Science and Engineering Forum

Simulation Studies of Surfactant Interactions with Carbopol-Type Ionic Polymer Gels

Aleksey Vishnyakov, Rutgers University, Piscataway, NJ

Andrei Potanin, Colgate-Palmolive, Piscataway, NJ

Alexander V. Neimark, Department of Chemical and Biochemical Engineering, Rutgers University, Piscataway, NJ

Surfactant-loaded polymeric gels are of great interest for the development of various consumer products. However, the physical-chemical mechanisms of gel stability and rheology are poorly understood, and the search for new formulations with improved properties is based mainly on empirical trial-and-error experimentation. We attempt to develop a molecular simulation approach aiming at a better understanding and quantitative predictions of morphological stability and rheology of gel-forming systems containing surfactants and cross-linked polymers like Carbopol. The multiscale approach includes molecular ab-initio optimization, thermodynamic modeling with Conductor-like Screening Model (COSMO), dynamics (MD) simulations of polymer-surfactant interactions at the atomic scale (2-5 nm), and dissipative particle dynamics (DPD) simulations at the mesoscale up to 50nm. We focus on gel-forming copolymers of highly hydrophilic polyacrylic acid and various hydrophobic compounds mixed with anionic surfactants p-(1-hexyl)heptyl benzenesulfonate (NaLAS), and sodium dodecylsulphate (SDS) and amino oxide (AO). SEM studies suggest that Carbopol forms a foam-like gel, which is destructed as the surfactant concentration increases. Our simulations shed light on this phenomenon. We demonstrate that surfactant molecules form micelles inside the polyacrylic acid structure without any appreciable interactions with the polymer. At the same time, surfactants interact intensively with hydrophobic linkers that may cause polymer network rapture. Using the DPD simulations, we show how the gel nanostructured morphology and the disjoining pressure in the polymer foam films depend on the chemical composition and surfactant concentration.

See more of this Session: Multi-Scale Modelling II
See more of this Group/Topical: Computational Molecular Science and Engineering Forum

Atomistic and Mesoscale Modelling of Poly(methyl methacrylate) in Order to Understand Its Structural and Dynamical Properties

Karl N. Kirschner, Computational Chemical Engineering Group, Fraunhofer Institute for Algorithms and Scientific Computing (SCAI), Sankt Augustin, Germany

Elena Algaer, AG Theoretische Physikalische Chemie, TU Darmstadt, Darmstadt, Germany

Hossein A. Karimi, AG Theoretische Physikalische Chemie, TU Darmstadt, Darmstadt, Germany

Florian Müller-Plathe, AG Theoretische Physikalische Chemie, TU Darmstadt, Darmstadt, Germany

Dirk Reith, Computational Chemical Engineering Group, Fraunhofer Institute for Algorithms and Scientific Computing (SCAI), Sankt Augustin, Germany

Poly(methyl methacrylate), also known as plexiglas or PMMA, is a common polymer that has a vast range of uses and applications. PMMA can occur in isotactic, syndiotactic, or atactic configuration, which affects its physical properties; relevant properties include its glass melting temperature, chain stiffness, miscibility, and surface activity. In this contribution, we present the development and application of a suitable model of PMMA for all three configurations.

Understanding the local potential energy surface and the dynamics of short polymer segments is an important step to utilize PMMA's properties in opening up future chemical modification strategies for new materials. Therefore, we show the results of molecular dynamics simulations performed on PMMA oligomers. The oligomers were composed of various residue lengths using a recently developed force field for saturated and unsaturated alcohols and esters. The simulations performed on the shortest polymer length, made up of three residues, was performed to determine the allowed conformational families of a PMMA triad. Quantum mechanics optimizations were subsequently performed to obtain a more reliable picture of the triad's potential energy surface.

In polymer science, however, the macroscopic properties greatly rely on the entropic contributions to the free energy of a specifically investigated system. Due to the broad range of time and length scales, not all questions can be dealt with at the atomistic level. That is because a detailed treatment of those degrees of freedom that govern small scales computationally prohibits the consideration of the slower modes that determine macroscopic properties. To overcome this problem, coarse graining to mesoscopic models is often applied. At this level, "superatoms" composed of up to ten atoms replace complete chemical repeat units. Therefore, we complement our atomistic study with a model transfer to the mesocale. We will show in detail how the mapping was executed and to which accuracy some structural and dynamical experimental properties could be matched for the various configurations.

See more of this Session: Multi-Scale Modelling II
See more of this Group/Topical: Computational Molecular Science and Engineering Forum

On the DPD Parameter Estimation From Atomistic / Quantum Mechanics Information

Maurizio Fermeglia, Molecular Simulation Engineering Laboratory, Department of Chemical Engineering, University of Trieste, Trieste, Italy

Paola Posocco, University of Trieste, Trieste, Italy

Sabrina Pricl, Molecular SImulation Engineering (MOSE) Laboratory - DICAMP, University of Trieste, Trieste, Italy

Johannes Fraaije, Leiden Institute of Chemistry, Soft Matter Chemistry, Gorlaeus Laboratories, University of Leiden, Leiden, Netherlands

Jan-Willem Handgraaf, University of Trieste, Trieste, Italy

Peter Deglmann, BASF SE, Ludwigshafen, Germany

Vandana Kurkal-Siebert, BASF SE, Ludwigshafen, Germany

Horst Weiss, BASF SE, Ludwigshafen, Germany

One of the major goals of computational material science is the rapid and accurate prediction of properties of new materials. In order to develop new materials and compositions with designed new properties, it is essential that these properties can be predicted before preparation, processing, and characterization.

Despite the tremendous advances made in the modeling of structural, thermal, mechanical and transport properties of materials at the macroscopic level (finite element (FE) analysis of complicated structures), there remains a tremendous uncertainty about how to predict many critical properties related to performance. The fundamental problem here is that these properties depend on the structure that the material exhibits at a length scale ranging from few to some dozens of nanometers, and this structure depends strongly on the interactions at atomistic scale.

To substantially advance the ability to design useful high performance materials, it is then essential that we insert the chemistry into the mesoscopic (MS) modeling. Currently, atomistic level simulations such as molecular dynamics (MD) or Monte Carlo (MC) techniques allows to predict the structure and properties for systems of considerably large number of atoms and time scales of the order of microseconds. Although this can lead to many relevant results in material design, many critical issues in materials design still require time and length scales far too large for practical MD/MC simulations.

Given these concepts, it is than necessary to carry out calculations for realistic time scales fast enough to be useful in design. This requires developing techniques useful to design engineers, by incorporating the methods and results of the lower scales (e.g., MD) to mesoscale simulations. One of the most reliable and used method for mesoscale simulation is the dissipative particle dynamics (DPD) in which the equation of motion is applied to agglomerates of matter (beads) that are subject to a soft potential. One of the main issue in applying DPD to real systems is the estimation of the parameters to be inserted in the soft potential. In particular the interaction parameter plays a relevant role in the morphology of the system.

In this work, we compare the results obtained from the application of two different methodologies for the estimation of DPD interactions parameters, namely (i) a recipe based on the calculation of interaction energies by atomistic molecular dynamics simulations and their mapping onto mesoscale energies, and (ii) the direct estimation of the DPD parameters by quantum chemistry calculations based on COSMO RS method.

The two methods will be compared using a test system of industrial relevance, for which complete and reliable experimental data are available, such as polystyrene-grafted silica nanoparticles dispensed into a polystyrene matrix.

Details on the 2 methodologies will be discussed, and results of the comparison will be reported both in terms of system morphologies obtained from DPD simulation and of macroscopic properties calculated by micro-FEM, using as input the nanostructured systems predicted at the mesoscale level.

See more of this Session: Multi-Scale Modelling II
See more of this Group/Topical: Computational Molecular Science and Engineering Forum

Combined Fluctuating Hydrodynamics and Molecular Dynamics Simulations

Jhih-Wei Chu, Chemical Engineering, University of California, Berkeley, Berkeley, CA

The goal of this work is to develop a systematic computational framework to determine the transport properties and equation of states of molecular fluids, especially those with spatial heterogeneity at the nano-scale. The essence of this framework is a direct mapping from the atomic representation of a molecular system to that of continuum mechanics, for which Landau's fluctuating hydrodynamics (FH) is employed. Since transport coefficients and equation of state are the input parameters for a FH representation, this mapping allows the determination of these properties directly from molecular dynamics (MD) simulations that use force fields. The results can also be used to improve molecular force fields in representing transport coefficients and equation of state. A key for this mapping to be meaningful is that the resulting thermal fluctuations of density and momentum fields of the FH and molecular scales need to be matched. By explicitly considering all relevant length-scales, we illustrate that this goal can be achieved. Furthermore, stable solvers for the FH equations are not yet available, especially when large fluctuations and transient flows are involved. We will also present a novel staggered discretization scheme that we developed for solving the FH equations with very small grid sizes (5-10 Å) and transient flows.

See more of this Session: Multi-Scale Modelling II
See more of this Group/Topical: Computational Molecular Science and Engineering Forum

AUTHOR INDEX

AUTHOR INDEX